TAKEN BY STORM, 1938

80TH ANNIVERSARY OF THE GREAT NEW ENGLAND HURRICANE

TAKEN BY STORM, 1938

A Social and Meteorological History of the Great New England Hurricane

LOURDES B. AVILÉS / 2nd edition

AMERICAN METEOROLOGICAL SOCIETY

Paperback ISBN: 978-1-944970-24-6
eISBN: 978-1-944970-25-3

Front cover photograph courtesy of the Boston Public Library, Leslie Jones Collection.

Published by the American Meteorological Society
45 Beacon Street, Boston, Massachusetts 02108

For more AMS Books, see http://bookstore.ametsoc.org.

The mission of the American Meteorological Society is to advance the atmospheric and related sciences, technologies, applications, and services for the benefit of society. Founded in 1919, the AMS has a membership of more than 13,000 and represents the premier scientific and professional society serving the atmospheric and related sciences. Additional information regarding society activities and membership can be found at www.ametsoc.org.

The Library of Congress has cataloged the hardcover edition as follows:

Avilés, Lourdes B., 1970–
Taken by storm 1938 : a social and meteorological history of the great New England hurricane / Lourdes B. Avilés. — 1st ed.
p. cm.
Includes bibliographical references and index.
Summary: "On September 21, 1938 the great New England hurricane hit the shores of New York and New England unannounced. The most powerful storm of the century, it changed everything, from the landscape and its inhabitants' lives, to Red Cross and Weather Bureau protocols, to the amount of Great Depression Relief New Englanders would receive, and the resulting pace of regional economic recovery"—Provided by publisher.
ISBN 978-1-878220-37-0 (hardcover)
1. Hurricanes—New England—History. 2. New England Hurricane, 1938. I. Title.
QC945.A95 2013
363.34'922097409043—dc23

2013017366

♲ Printed in the United States of America by King Printing Company. This book is printed on recycled paper with a minimum of 30% post-consumer waste.

CONTENTS

FOREWORD

by Ken Heideman, Director of Publications, American Meteorological Society, June 2013

Anyone who has ever experienced a hurricane does not forget it. And there are a few hurricanes that do not merely live on in the minds of individuals but leave lasting, searing impressions on communities, states, regions, and even nations. What are the characteristics of these rare, epic storms? Certainly, storms that inflict a large number of casualties and/or leave enormous economic losses, devastate infrastructure, and spread homelessness in their wake tend to make the grade. These impacts, horrific as they are, can be measured quantitatively. Such hurricanes leave a truly indelible mark in a larger context as well; they continue to have social, economic, and political impacts for months, years, and decades after the last of their devastating winds and torrential rainfall have ceased. There are a number of U.S. hurricanes that meet all of the requirements for enduring infamy. Katrina, Andrew, and Camille leap immediately to mind from the relatively recent past—since hurricanes have been named. Going back more than a century, the Galveston Hurricane of 1900, in which many thousands of people perished, is on this list. In between there were a number of major hurricanes, but one iconic storm stands out: the Great New England Hurricane of 1938. Other hurricanes have been stronger, lasted longer, and killed more people, but this one had no business hitting deep into the heart of New England after laying waste to Long Island, at least according to the meteorological wisdom of the time.

This storm is fascinating from a meteorological standpoint alone, but its place in history is assured because of the moment in time in which it occurred: a nexus of world events and social and technological changes that foreshadowed incredible advances, but at the same time left a large portion of the U.S. population particularly vulnerable to the unlikely occurrence of a major hurricane. Along the northeast coast of the United States in the early decades of the twentieth century, many affluent Americans built their dream vacation homes so close to the ocean they could smell the salt in the air from their front windows. Relatively poor but upwardly striving immigrants were arriving in great numbers and staking their claim to the American Dream, often settling near the coast to support their families and local economies by working in the fishing industry. The Great Depression was not yet over, but the promise of new scientific breakthroughs in science, communication, and manufacturing was palpable. And the winds of war, which would soon engulf the Unites States, were beginning to blow across Europe and beyond. One thing that was not on the minds of New Englanders in the midst of all this was hurricanes. They happened far away and to other people, in Florida and along the Gulf Coast. It had been so long since a hurricane had made a direct hit on the northeast United States that residents would have had no experience to draw on if they were warned that one was bearing down on them. As it happened, they didn't have the luxury of a warning on September 21, 1938. As the Great Hurricane moved northward off the Atlantic seaboard there were precious few surface observations of the storm, and the comprehensive network of upper-air observations in place today—so crucial to accurate forecasts of storm movement—did not exist. Moreover, it would be several decades before weather satellites provided constant surveillance of all weather systems. The science of meteorology was just starting to take huge leaps toward understanding the evolution and movement of tropical storms. All of these factors combined to catch Long Islanders and New Englanders completely by surprise; the region was dealt a lethal blow before they even knew what had hit them.

How different would the script have been if the Hurricane's path and intensity had been accurately predicted? Given the state of communications and weather forecasting at the time and how fast the storm was moving there would likely have been just a few hours of lead time under the best of circumstances. There was no interstate highway system, and evacuation orders, had they been issued and obeyed, would have had little practical effect. The lack of hurricane experience in the general population would likely have resulted in widespread apathy or paralysis even with foreknowledge of the impending storm. So while some lives surely would have been spared given advance warning, there still would have been many hundreds killed and multitudes injured. Houses along the shoreline still would have washed into the sea, the same storm surges would have brought the ocean into towns and cities, swollen rivers still would have jumped out of their banks, countless buildings and businesses would have been destroyed, and great swaths of forests would have been decimated. Of course an accurate forecast and advance warning would have helped, but by focusing primarily on those deficiencies (as is often done when this particular hurricane is remembered) as the reason for the ensuing disaster we are missing out on an essential lesson from The Great New England Hurricane. To wit, accurate forecasts are necessary but not sufficient to mitigate the devastation of weather disasters on the scale of this storm.

It took Hurricane Katrina in 2005 to snap all of us out of the fantasy that "if only forecasts were better" we could successfully withstand the worst blows that the atmosphere could inflict. Katrina was arguably one of the most accurately predicted major hurricanes in history: citizens, emergency personnel, and leaders at the local, regional, and federal levels knew for almost a week what was going to unfold from a weather standpoint based on widely disseminated forecasts from the National Hurricane Center. But we watched helplessly as the storm exposed vulnerabilities along social, economic, engineering, and political fault lines that transformed an event that should not have shocked us as something out of the realm of possibility—a hurricane making landfall along the Gulf Coast—into a catastrophe from which parts of New Orleans and the surrounding region have not yet fully recovered.

So why take such a careful look back at a hurricane that happened three-quarters of a century ago? To paraphrase Shakespeare in *The Tempest*, history is prologue. The question is not whether a storm like The Great New England Hurricane can happen again, but *when* will it happen again. Some contend that it has already happened, in the form of Superstorm Sandy in 2012. With all due respect to the fury of Sandy, I would argue that we are still waiting for the true encore performance of the 1938 storm that altered not only the course of history in New England but, quite literally, its very coastline as well. Next time, under the unblinking eye of weather satellites, we will not be ambushed, but we might be taken by storm in a different way. Indeed, the lure of breathtaking ocean views has not waned over time and the resident coastal population has swelled by many millions while sea levels have risen since the Great Hurricane, making the stakes even higher now.

In this carefully researched book, Lourdes Avilés traces the entire meteorological life cycle of the 1938 storm, from its inception in mid-September as a minor disturbance off of west Africa, to its survival and growth into a tropical storm and maturation into a hurricane as it crossed the Atlantic, and ultimately to its zenith as an intense hurricane and its transition to an extratropical system. It is a compelling read, to be sure, and addresses many questions and mysteries surrounding not only this particular storm but hurricanes in general. Above all, this is a one-of-a-kind reference work about a storm that still has a lot to tell us, even though it occurred 75 years ago in a different world, about how we can best survive the precarious dance in which humans and nature are continually engaged—if we choose to listen.

PREFACE TO THE 80TH ANNIVERSARY EDITION

The American Meteorological Society published the first edition of this book just a few days before the 75th anniversary of the Great New England Hurricane, attracting local, regional, and national attention. The landmark anniversary increased demand for stories, visuals, and more information about the storm, especially from expert sources. I did not anticipate the number of interviews I would be invited to do that week. I also had a variety of unique opportunities. I presented a talk as part of a three-night live-streaming event (additionally featuring historians, emergency managers, and others) organized by my fellow weather history enthusiast, WMUR Channel 9 meteorologist Kevin Skarupa. AMS posted an interview on YouTube between AMS Director of Publications Ken Heideman and me. I gave a presentation at the Mount Washington Observatory Weather Discovery Center to celebrate the opening of an exhibit sponsored by the Museum of the White Mountains, signed books at the Blue Hill Observatory anniversary event, and appeared via Skype on The Weather Channel. It was exciting and exhausting. Demand for talks about the 1938 Hurricane, however, continued past the anniversary.

Prior to publication of the book, through the anniversary events, and since (up to this writing), I have given 35 talks and presentations of various lengths and to various audiences, from general to technical, at historical societies, libraries, scientific meetings, and even in an 18th-century barn. I have crafted presentations with different focuses: precipitation observations, instrumentation, the role of women in storm relief activities, the storm as expressed in art, using historical storms to educate the public, using data and technical reports to tell a story, the early practices of the Weather Bureau—even birds! I curated the aforementioned museum exhibit about the effects of the storm in northern New England. I sat in on an episode (number 411) of the WeatherBrains podcast. I even found myself performing and self-recording a song I found in a Hurricane scrapbook (written by 4th grader Doris Deming; an excerpt of which appears on page 146).

The response has been overwhelmingly positive. I have received emails and handwritten notes from people thanking me for the book for reasons ranging from appreciation for new scholarly information to learning more about the extraordinary storm their parents had experienced. I am extremely thankful for how engaged so many people have been with my work. I am also humbly honored for receiving a Choice Award from Atmospheric Science Librarians International for a book of history.

The wonderful thing about giving these talks in the region that felt the storm's impacts is that you can have a mutual exchange of information. Local audiences are genuinely appreciative and interested in the story I am telling, but they also want to share their own, or the stories of their neighbors, friends, or family. Not many people who lived through the storm at the time are alive to talk about it now, but many people's lives today were changed by growing up with the stories (as well as by living in an era with forecasting and response standards that changed as a result of this storm). Through them, I continue to better understand how the science, numbers, and reports translate into the reality on the ground at the time and in the stories ever since.

Even the simplest stories are enlightening. I heard, for example, from a woman who, as a toddler in her stroller, was blown off the porch when the weather suddenly turned; her mother jumped down in dramatic fashion to rescue her. I heard from a man whose father stopped when the bridge across which he wanted to drive was under water; he got out of his car, approached the flooded bridge, and crouched down to take a closer look only to realize the bridge was no longer there. This was the year before the storyteller was born, meaning he might not have existed if his father did not make the decision to stop and check. I talked to another man who had pictures of his mother and friends playing on the fallen trees as if they were a jungle gym; I had read many times that children played in the storm wreckage, and I since added his photographs to my presentations. I also learned about interesting artifacts, environmental scars, and other signs left behind by the storm that can still be found if you know where and how to look. There are, for example, bowls and jewelry made with fallen wood, blown shingles on which information about the storm was written, elbow-shaped tree trunks in the forest, and huge depressions in the forest floor (the pits mentioned on page 205, which are much bigger than I imagined). There is also a pond (likely not the only one) where Forest Service–stamped pine logs, stored after the storm for their protection, still occasionally surface and float around for a while before sinking back down.

I sometimes come across new information when preparing to give my presentations, which I continually review and revise. I am amused by the irony that publishing the book led to the speaking opportunities that led me to discoveries I wish I could have included in the book. In this manner, I have learned additional background about some of the Hurricane relics originally featured in the book. I also obtained leads that helped me track down information I could not find before; for example, the fact that in 1938, 75% of the region was occupied by forest. Additionally, some definitions and practices related to hurricane forecasting and management have evolved in the years since publication, making those discussions in the book out of date.

And then there is what I consider a significant omission. I have been in possession of hundreds, if not thousands, of precipitation and other hydrologic observations that were gathered shortly after the storm by the United States Geological Survey (USGS) in a massive report I describe on pages 130–131 and 136–137. With the help of students, I had processed some of it; for example, river height data for the Connecticut River basin, which appears on page 137. From the start, I wanted to map the precipitation observations in their entirety, but there was not enough time to do so before publication. It would require several steps: digitizing, determining the latitude and longitude of every station (some of them quite obscure), and figuring out the appropriate tools to plot such data, as well as formatting the data so that it could be read by such tools. As far as I know, no one has done this since the team that wrote the USGS report gathered and hand-plotted the observations in 1940. I set it aside until I had more time. I offered it to a few students as a possible research project, but nobody took me up on it. Instead, our department's administrative assistant, Marsi Wisnuewski, together with her student workers, performed the initial data entry of the numbers, and it ended there for some time, until a very interested, resourceful, and fellowship-funded graduate student came my way. I am greatly in debt to Ms. Lauren Carter, who had the patience to check the digitized data, fill in the gaps, and perform the rest of the necessary tasks, resulting in a variety of beautiful GIS precipitation maps, some of which are featured in this edition.

One of these, together with many other updates, is included in the new Chapter 10. In order to accommodate this expansion, the original appendix (the report of the Weather Bureau to the Department of Agriculture found at the National Archives) is now available online only, which fortunately makes it possible to show the original report in all its glory, including photos of the original and all its attachments. Online supplements also include a variety of additional pictures and other visuals to supplement the information in Chapter 10. All online book materials can still be found at www.takenbystorm1938.com.

As I drafted the material for this update, and as I continue to learn new things about the Hurricane, my fascination (admittedly more mystical than scientific or historical) is once again rekindled.

Lourdes B. Avilés, P.h.D.
Professor of Meteorology
Plymouth, NH
September 2017

PREFACE TO THE FIRST EDITION

The hurricane-spawning machine of the tropical Atlantic was relatively quiet during the 1970s and 1980s. Still, during my childhood it seemed that every year at least one storm would threaten the island of Puerto Rico. Hurricanes, of course, can be bearers of great devastation, but to the little girl that I was, the promise of such an encounter was exciting. I wanted to go outside, feel the wind, and see the surf in the otherwise very calm Caribbean waters. I never put myself in danger, but I now realize that my fascination might have been a bit of a concern for my parents.

You can't evacuate an entire Caribbean island, but you can move people away from the areas of greatest risk. Our house, though not in a normally evacuated area, was a little too close to the ocean for comfort. This meant that every time a hurricane was expected, we would stay with my grandparents, as would the rest of my extended family. No school, playing with cousins, telling stories by candlelight: it all seemed more like a holiday or a party than sheltering from a storm. We were always safe at Mamita's house, a sturdy concrete house built on stilts and at higher ground than ours.

I was captivated by the stories my grandparents told about terrible *temporales* that blew in when they were young. I especially remember the stories about *El Huracán San Felipe*, which tragically killed more than 300 and left hundreds of thousands homeless when it devastated the island with category

5 winds. Also commonly known as the 1928 Okeechobee Hurricane, this is the same storm that drowned thousands when it later overflowed its namesake lake in southern Florida.

During those memorable sleepovers, I was the family member who stayed up all night with the battery-powered radio, eagerly awaiting updates on the whereabouts of a threatening storm. Of course, now I know how fortunate we were that big anticipation and big preparations never led to big devastation. Consequently, the memories of excitement and fascination prevailed over any justified fear of dangerous storms.

Many years after listening to my grandparents' stories and staying up listening to reports of approaching storms, the fascination is still with me, although maybe now it is mixed with a better sense of self-preservation. My life's journey has taken me far away from the Caribbean to northern New England, where I teach atmospheric sciences at Plymouth State University in New Hampshire. In this land of trees, waterfalls, and mountains, with four very distinctly beautiful seasons, winter storms are more of a year-to-year reality than any other type of hazardous weather. However, a few tropical hurricanes, which were such a memorable and defining part of my childhood, can—and do—make appearances here once in a while. And they can do it with a bang.

Even though my knowledge of hurricanes is much deeper than during my days in Puerto Rico, I strive to learn more, not just about the intricate and elusive science behind the storms, but also about past storms' historical context. It didn't take long for me to be drawn to the stories of the infamous 1938 Hurricane, the storm to which all New England hurricanes are compared. A call for Plymouth State's Faculty Week presentations combining one's area of expertise with topics of interest to the general faculty finally gave me an excuse to explore the storm in detail. I gathered everything I could get my hands on: books, reports, articles, newspapers, letters, pictures—you name it. Piece by piece, I put it all together into a story that combined the science and history of the Hurricane for local audiences. The next obvious step was to tell the story in more detail than a one-hour presentation allows. The approaching 75th anniversary of the storm and a teaching sabbatical combined to create the right conditions to plunge in headfirst and write this book!

A number of wonderful books cover the Great New England Hurricane of 1938, some recent and some from a long time ago. Shortly after the passage of the storm, several reports of damage, relief efforts, and meteorological analyses were quickly published as books or in scientific journals. Newspaper and magazine articles were also plentiful, and photo collections were immensely popular at the time. However, it was not until almost 40 years later that *A Wind to Shake the World*, a chronicle of the storm written mostly from first-hand accounts of survivors, was published by a reporter and survivor of the storm, Everett S. Allen. Almost 30 years after that, two other books filled with witness accounts and stories of the storm were published: *The Great Hurricane* by Cherie Burns and *Sudden Sea* by R. A. Scotti. A third book, *The Hurricane of 1938* by Aram Goudsouzian, contains great information about the historical times in which the storm occurred. These books are in print as I write this, but after some digging I also came across a few obscure, out-of-print publications. Still, books on the subject with a scientific perspective have not been as plentiful. Only one, published 25 years ago for the 50th anniversary of the storm, by William E. Minsinger, then president of the Blue Hill Observatory (where the strongest wind gust was measured during the storm), comes to mind. The book contains a collection of scientific articles and accounts, summarizing the meteorology of the Hurricane and providing a survey of historical New England hurricanes. Through these books, together with various scientific journal articles, reports, visits to historical archives, and many other resources, I acquired a wealth of information and complementary perspectives that have shaped my understanding of many different facets of the storm. From a more general perspective, one of my original inspirations was Kerry Emanuel's *Divine Wind*, which combines hurricane science chapters written for a general audience with historical chapters.

My goal has not been to retell the story that has already been told, although there has to be some of that too, but to take a somewhat interdisciplinary approach to weaving together different aspects—different stories—of the 1938 Hurricane. This includes what happened before, during, and after the event, in the context of the meteorological history of the storm and its associated destruction and devastation; casualties, survival, and recovery in the affected population; environmental and geological changes caused by the storm; the science of hurricanes and of early-20th-century meteorology; and, finally, the added perspective of other intense hurricanes that have affected and no doubt will again affect the region.

A couple of caveats are in order for those who wish to follow this journey with me. Hurricane science can be very complicated, and much of it is beyond the scope of this book. I present as much of the science as is possible and relevant, while striving to make the information suitable for a general audience. Additionally, the reader will note that the chronology will not be linear since both the history that led to the events of the storm as well as how they fit in with our current knowledge and perspectives are important to properly tell the story of the Hurricane. Finally, even though most of the events occurred decades ago, our knowledge of the atmosphere and the way in which we deal with hurricane preparedness, monitoring, forecasting, and warning continue to evolve. New pieces of evidence continue to surface, new technology constantly emerges, new warning definitions and practices are put in place, and new data analyses are performed, meaning that some information in this book will continue to evolve for years to come.

In addition to the content in the book, you will find a depository of supplementary materials at a companion website, *More on the Science and History of the Great New England Hurricane* at www.takenbystorm1938.com. The site contains scientific and historical resources directly and indirectly related to the Hurricane as well as further data and calculations. Additionally, it contains a full list of sources, links to online resources, electronic versions of some historical documents and books, and high-resolution versions of some of the images. This is also where any new sources or updated information will reside.

It took a great deal of effort to complete this project, and I could not have done it all on my own. First of all, I am not sure that I would have written this book yet if it were not for a community education talk that I was asked to present at the Taylor Retirement Community in Laconia, New Hampshire. Both preparing that talk and then delivering it to the best audience I have ever had helped me find the courage to take on such a large project. I also appreciate the support I received and continue to receive from Plymouth State University and its meteorology program, which has allowed me to pursue my research interests within a very busy schedule. I am additionally indebted to several individuals that have provided me with pieces of information, have helped me uncover resources, or have helped in other big and small ways. I would like to give them all a little shout-out: Hilda Avilés, Scott Bailey, Katie Rose Boissonneault, A. J. Coppola, Gregg Champlin, Terri Dautcher, Neal Dorst, Kerry Emanuel, Jim Fleming, Richard Giard, Mark Green, Andrew Hagen, David Heeps, Bob Henson, Brendon Hoch, Eric Hoffman, John Krueckeberg, Chris Landsea, Sam Lillo, Stephen Long, Derek Mallia, Don McCasland, Alyssa Mira, William Minsinger, Erika Moore, Jinny Nathans, Rebecca Noel, Erin Norris, Wendy Palmquist, Jared Rennie, Cort Scholten, David Schultz, Donna Strahan, Alice Staples, Samuel Tolley, Marsi Wisniewski, and a few others who I am either failing to recall or who have preferred to stay anonymous. I would, of course, also like to thank the American Meteorological Society for the opportunity to share my research and its resulting stories; Sarah Jane Shangraw, Ken Heideman, and

Beth Dayton have been most enthusiastic, supportive, and helpful throughout this process. Last and certainly not least, I would like to thank my family for their encouragement and for suffering my absence, and especially my husband, Dan Bramer, for his continuous and unconditional support as well as his excellent technical help.

As atmospheric scientists, we classify storms and weather systems out of our need to understand, but sometimes they do not easily fit into a category. In writing this book I have learned to appreciate how the same is true when studying the events of the past. We try to fit our understanding of everything that happened into simple descriptions, but the truth is always more complicated—and more interesting! The Great New England Hurricane of 1938, its historical context, and the science and meteorology behind it, all of which have been the subject of my research during the past few years, weave an intriguing and compelling account of an event that is also a cautionary tale for present and future generations.

Lourdes B. Avilés, Ph.D.
Professor of Meteorology
Plymouth, New Hampshire
June 2013

PROLOGUE

On Wednesday, September 21, 1938, the forecast called for heavy rain in the afternoon for most of New England, but rain was not big news these days. It had been a rainy summer and it had even rained enough during the previous week to flood rivers and streams throughout the region. The unfriendly weather was a favorite topic of complaint that year, but on that particular morning, for those left behind as the summer crowds dwindled, life went on as normal. Parts of Long Island and southern New England even woke up to a sunny morning, and some folks who didn't have to go to work or school hastily put together a picnic or a beach outing before the predicted return of the rain. Little did anyone suspect that a storm coming from far away in the tropics was about to bring unimaginable destruction. A massive hurricane was going to arrive unannounced, and it would kill several hundreds that day, many of whom would drown within the wall of water that came inland with its sudden landfall.

No major hurricane had affected the region in more than a generation, certainly none that was remembered in the minds of those in charge of forecasting on that day. Intense hurricanes simply did not happen in New England—or so it was assumed, and with good reason. Atlantic hurricanes, which develop in the tropics and usually track westward and then northward around the Bermuda High, often weaken and fizzle out well offshore, somewhere over the North Atlantic Ocean. Those that do not take this early turn avoiding the North American continent can maintain their intensity more successfully in the warmer tropical and subtropical waters, but normally hit land much farther south, somewhere on the coast of the Gulf of Mexico, Florida, or south of Cape Hatteras, North Carolina. The right combination of atmospheric conditions needs to be present for a particularly intense hurricane to take a path straight toward the northeastern United States while maintaining its strength; but the tools needed to make the appropriate meteorological observations of such conditions—and to formulate an accurate forecast based on those observations—were only just becoming available.

Unknown to forecasters in 1938, all the right conditions would in fact converge on this day, and a massive, powerful, and unprecedentedly fast-moving hurricane was heading directly for Long Island and New England. The Hurricane resulted in great devastation that spanned from the coast deep into the forest and mountains of interior New England.

The impacts were unprecedented. How did this happen? How could such a storm shatter a region then commonly believed to be immune to strong hurricanes? What caused its remarkable forward speed and its seemingly unusual path? Why didn't anyone predict its arrival? How did the Weather Bureau explain the lack of adequate warning? Was it even possible for forecasters to have done any better? How did the people and the environment recover? What will happen when such a storm occurs again in the future? What lessons have been learned from the storm? These and many other questions warrant a careful look at the Great New England Hurricane of 1938, one of the most devastating in the history of the United States.

PART I
OVERVIEW

1

THE GREAT HURRICANE OF 1938

The storm, the science, and the history

Even during the Depression years, New England and Long Island were summer playgrounds to the rich; many affluent city dwellers owned beach- and lake-front cottages. The summer of 1938, however, did not boast great weather for outdoor enthusiasts, and as usual, by September, most of the seasonal residents had left their coastal getaways to return to the city.

Autumn was on its way. It had been a wet, gray summer, and since September 17 the weather had been extraordinarily rainy. Several inches had fallen over much of the area, softening the ground and leaving streams and rivers near or already above flood stage. On September 21, after what must have seemed like unending overcast skies, southern New England finally awoke to beautiful sunny skies. Just to the south, Long Island awoke to cloudy, steamy conditions, but no rain. Residents went about their business, going to school or to work, running errands, or simply enjoying some leisurely time.

Meanwhile, the political climate overseas had the whole world on edge. Adolph Hitler's Germany had recently invaded Austria and now had set its sights on the German-speaking Sudetenland region of what was then Czechoslovakia.[1] Hitler's expansionist behavior and disregard for post–World War I agreements had become increasingly obvious. By early to mid-September, newspapers were bursting with news of the British government's efforts to

The shores of Long Island and southern New England have served as a summer playground for many. (July 1958 *Weather Bureau Topics* cover art)

avoid war.[2] Prime Minister Neville Chamberlain led talks and negotiations, which would culminate with the signing of the Munich Agreement and his infamous proclamation of "peace for our time" on September 30. The agreement, signed during a conference lacking Czech and Slovak representation, permitted Germany's annexation of the Sudetenland in exchange for Hitler's promise to cease any further expansion activities. Less than a year later, on September 1, 1939, Germany would invade Poland, breaking the agreement and giving rise to a declaration of war by Britain and France, thus beginning World War II. For now, on September 21, 1938, as heads were turned toward the political storm clouds gathering in Europe, a destructive force of a different kind was approaching the northeastern seaboard without warning.

In the days preceding the Great New England Hurricane, hidden among the news of European events, were a few blurbs about a storm heading for Florida. On the morning of the Hurricane's eventual landfall in the northeast United States, news highlighted the good fortune that the storm had turned away from Florida's coast. Meanwhile, radio stations devoted their time broadcasting coverage of the European political situation, some via correspondents on location in Europe. In fact, a Weather Bureau report to the Secretary of Agriculture just two weeks after the storm mentioned the following:

> The unfortunate fact that one of the powerful New York radio stations upon which so many residents of Long Island, Connecticut and adjacent areas depend for their weather information failed to broadcast the important storm information was probably due to the popular interest in broadcasts of information regarding the threatened war in Europe, result[ing] in the failure of thousands of residents in the affected area to receive our warnings, in time or perhaps not at all in many cases.[3]

Equally telling is that on the morning after the Hurricane, the front page of the *New York Times* gave two stories equal billing: "Hurricane Sweeps Coast . . ." and "Czechoslovakia Gives Up. . . ."[4]

At first light on September 21, to the untrained eye there was no sign in the sky of the upcoming disaster. While a trained observer might have taken note of the unnaturally calm weather coupled with gradually enhanced oceanic swell over the last few days, or the increasing frequency of crashing waves on the shore, or even the lack of birds overhead, a tropical hurricane was the furthest thing from the minds of the citizens of Long Island and New England.[5,6]

What's in a New England Hurricane?

Hurricanes in the New World

Large, intense, rotating storms develop over tropical and subtropical oceans around the world. In the Atlantic and Eastern Pacific Oceans, these remarkable storms are called hurricanes. The word originates from the Spanish *huracán*, which is in turn a phonetic interpretation, as heard by early Spanish settlers, of the name called out by the New World natives during extraordinarily destructive storms.[7] By most accounts, a weather-controlling mythological deity named Jurakan (hoo-rah-kahn) unleashed its fury in the form of the terrible storms that plagued what today are the Caribbean islands. The origin of the legend, usually attributed to the Taínos (who populated a large portion of the region during the Spanish colonization), most likely was rooted in a similar deity within Mayan mythology.[8] Not surprisingly, the Taínos feared Jurakan as an evil spirit, and they attempted to drive it away with shouts and chants and ceremonies that would have surely caught the Spaniards' attention.[9]

Yet, the word *huracán/hurricane* didn't always mean exactly what it does today. Before settling into the current form, many early spelling variations of the resulting Spanish word and its English translation were used (e.g., *furicanes, vracans, huricans, herricanos*).[10] Additionally, the informal designation of "hurricane" was initially applied to different types of storminess, referring more to the extreme wind intensity than to the storms themselves. Today, a hurricane is very specifically defined as a tropical cyclone with maximum sustained surface winds of at least 74 mph.[11,12] "Sustained winds" refer to measured or estimated one-minute-averaged speed at a standard height of 10 meters (33 feet) above the ground.[13] They are not the same as the instantaneous gusts of wind, which can be stronger and extremely variable. The same storms are given a variety of names throughout the world. For example, typhoon is used in the Northwest Pacific Ocean and severe cyclonic storm in the North Indian Ocean.[14] Tropical cyclone, however, is the general name used internationally to describe all large-scale, rotating, tropical low pressure systems not accompanied by fronts—even those of lesser intensity than hurricanes.[15] Anywhere from a handful to more than 20 tropical cyclones, including hurricanes and their less intense versions, tropical storms (tropical cyclones with maximum sustained winds of 39–73 mph), form over Atlantic waters every year. Given the path taken by many these storms, it is very likely that some of the same hurricanes experienced by Caribbean natives and Spanish settlers went on to affect the land of what later became known as New England.[16] It is also very likely that these same storms were the cause of great devastation throughout the area, although affecting a much smaller population than existed early in the twentieth century.

New England is no stranger to storms of tropical origin. Every year or two, tropical cyclone remnants bring with them some combination of heavy rain and high surf, depending on their specific tracks. Much less often (every decade or two) a fully formed hurricane makes direct landfall in the region.[17,18] A hurricane as devastating as that of 1938 skips generations, occurring only about once every century or even less often. Only the Great Colonial Hurricane of 1635 and the Great September Gale of 1815 are believed to have caused as severe coastal flooding and as large an area of devastation. There have been a few other noteworthy New England hurricanes since then, namely the 1944 Great Atlantic Hurricane, Carol (1954), Donna (1960), Gloria (1985), Bob (1991), and Irene (2011), but none like the one that arrived on that fateful Wednesday in September 1938 (see Chapter 9 for more information about other historical and recent New England hurricanes).

Tropical versus Extratropical Cyclones

Hurricanes differ in many ways from seemingly similar storms that affect the northeastern United States much more frequently. The familiar "lows" that cross the United States repeatedly throughout the year—often drawn on weather maps—are capable of bringing with them a great variety of hazardous and interesting weather. Everything from snow to rain, sleet to hail, and blizzards to tornadoes can accompany these systems at one time or another (depending on the time of year and the specific meteorological conditions).[19] Some of them eventually develop into the infamous coastal storms commonly known as nor'easters (named for the direction from which winds blow during such storms), perpetrators of much mayhem as well as excitement for New England residents. Meteorologists often refer to these general types of lows as extratropical cyclones to distinguish them from the much different tropical cyclones that include hurricanes. This distinction has not always been made. During the early 19th century, these events were simply referred to as *storms*; but even then it was becoming clear that, at least in New England, "northeast storms" (most of which would have been nor'easters) seemed to cause different types of weather than "southeast storms" (many of which we now know were in fact hurricanes).[20,21] Both are low pressure systems that can bring strong winds, high waves, and flooding to the region, but that is where the similarities end. Even the most intense extratropical cyclone is barely as strong as a minimum-intensity hurricane. Additionally, extratropical cyclones tend to be hundreds of miles wider than their tropical counterparts. The differences, however, go well beyond a tropical or non-tropical origin, the strength of the winds, or the size. Significant structural and energetic differences also distinguish the two types of storms.

Unlike hurricanes, extratropical cyclones do not manifest in a symmetrical circular arrangement and normally do not contain an eye. In their mature form, their cloud structure resembles a comma. This appearance is intimately related to the temperature and moisture contrast between what meteorologists call the "air masses" surrounding the storm. An air mass is simply a very large volume of air, roughly one to three thousand miles wide, with homogeneous temperature and moisture characteristics. The boundary where two such air masses of contrasting characteristics meet is known as a front.[22] With colder and usually drier air to the north and warmer and usually more humid air to the south, extratropical cyclones are always accompanied by fronts. As the (less dense) warm and moist air is forced to rise over the advancing colder air, clouds, precipitation, and sometimes thunderstorms develop along the often north-south-oriented cold front.[23] Additionally, the clouds partially wrap over the often east-west-oriented warm front, around the center of low pressure in counterclockwise fashion (in the Northern Hemisphere). The result is the formation of the characteristic comma-shaped cloud arrangement that can be seen on satellite images. The temperature contrast between air masses is also what provides much of their energy. Potential energy—which exists because of the difference in density between the adjacent cold air and warm air—is converted into kinetic or motion energy in the form of stormy winds.[24] In contrast, tropical disturbances have no fronts; in their purest form, they are fully embedded in warm, humid, tropical air. Their energy is ultimately provided by the warm ocean surface, which together with a favorable environment can set in motion a complicated—and still not fully understood—self-sustaining mechanism that can lead to a tropical storm and, in the extreme, a hurricane (see Chapters 3 and 4 for more information about hurricane formation and intensification).

Clearly, hurricanes are very different from extratropical cyclones. However, as happens in nature, atmospheric phenomena do not always fully fit in one category or another. The line between the two is sometimes blurred and there can be hybrid storms with characteristics of both. These "misfits" are usually dubbed subtropical storms, but they do not last too long and tend to soon transition into purely tropical systems.[25] It is also not unusual, if suitable meteorological conditions become available, for hurricanes to undergo an extratropical transition instead of simply dissipating after they move away from the warm tropical oceans where they originated. (Hurricane Sandy [2012] was an interesting hybrid storm for a while due to an incomplete extratropical transition; see Chapter 7 for more information about the extratropical transition of tropical cyclones.) This was the case with the 1938 Hurricane; as it sped northward and then devastated New England, it transitioned into an extratropical storm. The possibility of this type of transformation had been suspected by a handful of western Pacific typhoon experts during the very early years of the 20th century (see Chapter 5), but the concept was more of a theoretical assumption at the time; they lacked sufficient data to test the theory. A very careful postanalysis of the Great New England Hurricane was the first study to unexpectedly show the presence of different air masses and the most likely transition of energy source, providing the first observational evidence as well as a modern and detailed meteorological analysis of the extratropical transition of hurricanes.[26] It is not possible, however, to determine exactly when the transition was completed or if the storm was still technically a hurricane when it hit Long Island in the early afternoon. (It is worth clarifying that the strength of the winds was most definitely of major hurricane strength during landfall, but a storm must also have the right kind of structure, as described above, in order to be a true hurricane.) The best that can be said is that it was most likely transitioning at that time, since the storm was without doubt fully extratropical as it moved through northern New England just a few hours later.[27]

Naming Storms

The 1938 Hurricane was not assigned a proper name like other modern-day hurricanes—Katrina and Andrew might come to mind as particularly devastating named hurricanes. During the subsequent months and years after the storm, the event was referred to by different names. In its November 1938 report, the New Hampshire Disaster Emergency Committee, for example, referred to the event as the Flood and Gale of September 1938, whereas in its February 1943 report, the U.S. Forest Service simply named it the New England Hurricane of September 21, 1938. The earliest known naming practice originated in the Caribbean Spanish colonies. For hundreds of years, since the early 1500s, settlers had given most hurricanes the Spanish name of a saint whose feast was celebrated on the day of the storm's landfall according to the Catholic calendar. The practice, however, provided room for confusion because a storm could be known by two different names on two nearby islands experiencing its landfall on two different days. The official naming of Atlantic storms, motivated by the need to distinguish them from one another, did not start until the 1950s, first using a military phonetic alphabet (Able, Baker, Charlie, Dog, etc.) and then exclusively female names.[28] The practice of using saints' names continued in parallel to the official naming but was abandoned after 1960, when Hurricane Donna (which coincidentally later went on to affect New England) was also locally known in Puerto Rico as Huracán San Lorenzo.[29] The modern naming system used for Atlantic hurricanes, with six rotating alphabetical lists of alternating male and female names in English, Spanish, and French, did not come into use until 1979.[30] The Great New England Hurricane of 1938, also well known as The Long Island Express, thus earned its name by virtue of the profound and multiple effects it had on the people and the environment of the region.

The Great Hurricane by the Numbers

Historical hurricane records for the North Atlantic Ocean, Caribbean Sea, and Gulf of Mexico extend back more than a century and a half. The North Atlantic Hurricane Database (HURDAT) officially holds these records. This comprehensive compilation was first put together during the 1960s in support of the Apollo space program.[31] The database, which consists of thousands of lines of coded data, contains estimates of maximum sustained winds, minimum central sea level pressure, and latitude/longitude positions at six-hour intervals for all known hurricanes and tropical storms since 1851. Since its initial creation, the database has been updated every year with what is known as the "best-track," consisting of quality-checked position, intensity, and other data resulting from post analyses by the National Hurricane Center (NHC) performed for all the year's storms once the hurricane season is over.[32] The records originally went back to 1886, but in 2002 they were extended even further back to 1851 after extensive studies of historical documents provided thousands of new pieces of information.[33,34] HURDAT has served as the main data source for many climatological hurricane studies, but the tools available to observe and analyze hurricanes have changed significantly throughout the years, which means that artificial errors have been introduced into the raw data and one must be careful when interpreting any type of analysis or statistics obtained. To make up for this artificial unevenness as much as possible, a reanalysis of all available information related to all known hurricanes and tropical storms is systematically being performed. Meteorological reanalysis projects, which consist of the processing of data obtained over long periods of time by using consistent analysis methods, are a common way to improve the quality of datasets that can be used for climatological studies (see Chapter 2 for more information about the Hurricane Reanalysis Project). HURDAT itself has also very recently undergone an upgrade in content and format. The resulting improved database, HURDAT2, was introduced in November 2012; it contains additional information such as landfalling data and wind radii. It also includes the weakest of tropical cyclones, tropical depressions (with sustained winds below 39 mph), which were not part of the original database if they did not intensify into at least a tropical storm.[35] In this book, the Atlantic Hurricane Database will be referred to as HURDAT/2, unless referring specifically to the original or the updated version.

The reanalysis for the 1938 hurricane season was finalized in December 2012. According to the updated data, nine storms formed during that year, seven of which hit land somewhere on the coastal United States, the Bahamas, a Caribbean island, or Central America. Based on their maximum intensity, five were classified as tropical storms and four as hurricanes. There might also have been a small number of tropical depressions, but as mentioned above, these would not have been included in the original database, and in reality they would have been much easier to miss in the earlier years. The HURDAT2 upgrade also does not contain any additional nonintensifying tropical depressions for the year. The Great Hurricane was the sixth storm and the fourth hurricane to form. It was only the second to make

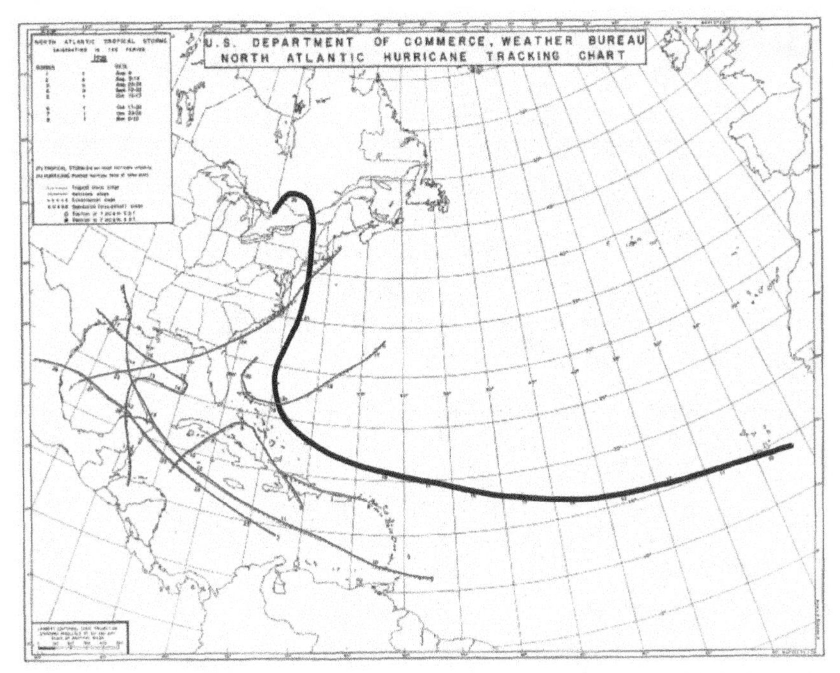

Before the 2012 reanalysis of the 1938 hurricane season, all known storms for the year appearing in the Atlantic Hurricane Database had made landfall somewhere in the United States, Central America, or Caribbean coasts. Two more storms that did not affect land (not shown) have now been added, and one (originally #7) has been taken off the list. The Great Hurricane (highlighted) was the only one to develop off the coast of Africa and the only one to affect land with major hurricane intensity. (Weather Bureau Chart based on the original HURDAT)

direct landfall over the contiguous United States and the only major hurricane to form that year, reaching a category 5 on the Saffir–Simpson Scale (described below) during its maximum intensity over the ocean.

The Saffir–Simpson Scale did not exist in 1938. It was developed in the 1970s by wind engineer Herbert S. Saffir and meteorologist Robert H. Simpson (who was also the director of the NHC from 1968 to 1973). Originally named The Hurricane Disaster Potential Scale, it has been used since then to assign all hurricanes, including all known historical hurricanes, a category from 1 to 5 based on their maximum sustained winds.[36] As originally conceived and used for almost four decades, the scale had included maximum storm surge (the rising of the sea height due to a hurricane's landfall) and minimum atmospheric pressure at the storm's center in association with each category. Over the years, however, it has become clear that, while

SAFFIR–SIMPSON HURRICANE WIND SCALE

The Saffir–Simpson Scale is a 1 to 5 rating of a hurricane according to its sustained wind speeds, which directly relate to potential property damage. Hurricanes reaching category 3 and higher are considered major because of their potential for significant loss of life and damage, but category 1 and 2 storms are still dangerous. Damages due to other storm hazards, such as storm surge and inland flooding, are not included in this rating.

Saffir–Simpson Hurricane Wind Scale

Category	Sustained Winds (mph)	Damage Due to Hurricane Winds
1	74–95	Very dangerous winds will produce some damage.
2	96–110	Extremely dangerous winds will cause extensive damage.
3 (major)	111–129	Devastating damage will occur.
4 (major)	130–156	Catastrophic damage will occur.
5 (major)	157 or higher	Catastrophic damage will occur.

NOTE: For more information and related links, see www.nhc.noaa.gov/aboutsshws.php.

representative on average, the specified surge and pressure could be misleading for any individual storm. The actual central pressure that accompanies the observed maximum winds in the eyewall of a hurricane is affected by the size of the storm. Of even more significance is the fact that the actual storm surge experienced during landfall is profoundly affected by factors such as the shape and depth of the coastline, the angle at which the storm approaches, and once again, the storm size.[37] Because of the large potential variability and the public misconceptions that this variability could cause, together with their public safety implications, the scale was redefined in 2010 as the Saffir–Simpson Hurricane Wind Scale.[38] The wind speed range corresponding to each category was kept the same, but all references to storm surge and pressure have been removed. A hurricane with sustained winds of 74 to 95 mph is still classified as category 1, as it was with the original scale; a major hurricane, defined as category 3 or higher, still has wind speeds greater than 110 mph, for example. The revised scale now additionally provides examples of the types of damage and impacts directly associated with the intensity of the wind indicated for each category.

TABLE 1.1. The Great New England Hurricane of 1938 by the Numbers

Formation/Dissipation	September 10 / September 23
Maximum Intensity	SS Category 5 140 kt (160 mph)
Lowest Pressure	938 hPa[a]
Landfall	September 21 Long Island—2:45 PM EST Connecticut—3:40 PM EST
Landfall Intensity	SS Category 3 105 kt (120 mph), 941 hPa
Death Toll	600–800
Approximate Cost	$250–450 million 1938 USD $4–$7 billion 2012 USD

SOURCE: Estimated and from various sources

[a] One hectopascal (hPa) equals 100 pascals. A pascal (Pa) is the international unit of pressure, equal to a Newton per square meter or a kilogram per meter second squared.

At landfall, the 1938 Hurricane had the strength of a major hurricane, specifically category 3 (see Table 1.1), with winds of 120 and 115 mph experienced as it came onto Long Island and Connecticut, respectively. It should be noted that this is as strong as it gets in this region; no hurricane stronger than a category 3 has ever been observed or inferred to have made landfall in New England. (The strongest winds ever achieved by the storm, on the other hand, are estimated to have been probably as high as 160 mph, or category 5, at its most intense phase just east of the Bahamas.) This storm provides a perfect example of why it was necessary to disassociate possible storm surge heights from the Saffir–Simpson classification. According to the original scale, a category 3 hurricane as the 1938 Hurricane was classified during landfall would produce a storm surge (coastal water rise) 8 to 10 feet (above normal). This was the case in some portions of the affected region, but there were also a few locations where the shape of the coastline significantly enhanced the surge, producing maximum water heights of over 20 feet above mean low water (see Chapter 6 for more specifics on the Hurricane's storm surge).[39] Such water heights (if mostly from the storm surge rather than the normal astronomical tides) would have corresponded to a category 5 hurricane in the old formulation of the scale. There are several other similar examples, making the separation of hurricane intensity classification and the forecast of other potential impacts caused by a hurricane an important improvement in keeping the coastal population safer and properly informed.[40] The development of a storm-surge-specific scale has been advocated by some scientists, but at this time the NHC has decided against it and believes that the best approach is to clearly convey to the public the depth of inundation expected at their location.[41]

The 1938 New England Hurricane makes appearances in lists of both the deadliest and the costliest hurricanes to have affected the United States.[42] Hundreds died, many by drowning in the unexpected storm surge. The generally stated figure is 682 fatalities, but reports vary from the low to the high hundreds and, as is the case with many natural disasters, the exact number will never be known. The fact that most of the summer residents had left the region toward the end of the month meant that, mercifully, this tragically large number of casualties was much lower than it could have been.

The same is true for reports of estimated cost, which roughly range from $250 to $450 million in 1938 dollars. A large amount of property damage contributed to this cost, with approximately 20,000 buildings (including homes, summer cottages, barns, and other types of structures), 26,000 cars, and 5,000 boats destroyed or damaged.[43] Unfortunately, a great majority of the losses (about 95 percent) resulting from damaged property were uninsured, meaning that much of the cost was endured by the affected individuals—although a portion was subsidized by personal loans and a handful of federal programs (see Chapter 8 for more information on the role of the various federal programs in the relief and recovery efforts after the Hurricane).[44] Massive infrastructural damage also contributed to the cost, with 20,000 miles of electric and telephone cables downed by the storm, and thousands of road, railroad, and bridge washouts having to be repaired. Just adjusting for inflation, the 1938 numbers would translate to approximately $4 to $7 billion in 2012 dollars, but the same storm occurring today would likely have a much higher cost.[45] The calculation of such a cost is complicated; besides inflation, changes in other factors such as wealth and population can also have a great impact on the cost of a specific storm occurring at a certain time in history.[46] Additionally, a small difference in the track can have a significant impact on the cost. The worst of the damage is normally caused by the portion of the eyewall to the right of the storm's motion—in other words, to the east of the eye for a northward-moving storm like the 1938 Hurricane. For this specific storm, New York City was, though just barely, on the "good" side to the west, avoiding the storm surge (but not the large waves) and the strongest winds (but not the heavy rain). If a similar storm made landfall just a few tens of miles to the west, the potential maximum cost could very well top $100 billion, in the same ballpark as for Hurricane Katrina—which is as of now the costliest storm of all time.[47,48,49] Sandy (2012) took such a worst-case scenario track and did cause a huge amount of infrastructural damage in New York and New Jersey, but it still fell short in strength compared to the 1938 Hurricane, or its damages would have been much worse (see Chapter 9 for more about the calculation of the cost of a 1938-Hurricane-like storm in modern times and about Hurricane Sandy).

The 1938 Hurricane as a Catalyst for Change

The statistics are remarkable (see Table 1.2 for additional information), but the effects of the storm go much deeper than numbers. Lives, homes, and livelihoods were lost. Communities were deeply altered. With roads and bridges washed out, railroad tracks destroyed, and telephone and telegraph services down for weeks or longer, towns lost communication with the rest of the country, which had little idea of the magnitude of the disaster. Also, with power lines downed, electricity (which had only become widely available to the general population within the previous couple of decades) was out in some areas for as long as a month. A region self-considered to be modern and sophisticated was violently thrown back to the times of kerosene lamps, candles, bellied stoves, and fireplaces.[50]

For businesses and the economy, the Hurricane could be said to have come through as a double-edged sword. Most obviously, it struck the final blow to many mills and factories that had been struggling in the outskirts of the Great Depression. At the same time, it stimulated an avalanche of economic growth that eased the effects of the Depression for many. Economic renewal is not an unusual side effect of a massive natural disaster, with similar effects having been attributed, for example, to the 1906 San Francisco Earthquake, the 1900 Galveston Hurricane, and the Great Chicago Fire of 1871.[51] Rebuilding the region was an enormous task, and any able-bodied person who wanted a job could have one. Building materials such as shingles, nails, and hammers, as well as logging tools such as portable sawmills were in great demand. Local suppliers were not able to fill the many demands, and New England became a prime destination for goods from around the country. It would not be until the massive industrial mobilization needed for World War II that the United States would fully come out of the Depression, but the Hurricane certainly alleviated its effects for the region. An interesting example comes from the budding airline industry, which was the recipient of a permanent boost in business when train service between Boston and New York was disrupted for weeks. Many would-be rail passengers, mostly business travelers, had been hesitant to fly in the past and now had no choice but to discover the convenience and speed of flying.[52]

TABLE 1.2. More 1938 New England Hurricane Numbers

Additional Storm Statistics	186 mph—strongest wind measured over land	Highest wind gust at Blue Hill Observatory
	50 mph—fastest forward motion	(estimated possible fastest motion)
	30 miles—radius of strongest winds	(RMW—radius of maximum wind)
	200 miles—width of damage path	Worst damage 100 miles east of center
	25 feet—highest water level	Storm surge plus astronomical tide above average low water levels
	30- to 50-foot waves during landfall	Above storm tide
	27.94 inches—lowest pressure measured at a land station	Corresponds to approximately 943.5 hPa
	1 hour—maximum duration of calm center	50 miles—maximum estimated eye width
Personal Losses	700 deaths	Highest estimate
	1,800 injuries	Highest estimate
	93,122 families reporting losses	
	19,608 families requesting emergency help	From the Red Cross
	842 in hospitals	
	15,107 in Red Cross shelters	
	6,000 received first aid	
	11,000 received inoculations	Typhoid serum
Property Losses	19,639 buildings	15,139 damaged and 4,500 destroyed, including year-round and summer homes, barns, and other buildings (Red Cross numbers, other higher estimates exist)
	5,974 boats	Including 3,369 damaged and 2,605 destroyed
	26,000 cars	Estimate from insurance claims
	95% of property losses uninsured	
Infrastructure Damages	7–14 days railroads halted	
	10,000 transportation workers	Fixing 1,000 road and railroad washouts and 100 bridges
	100,000 WPA (Works Public Administration) workers	Doing everything from rescues, cleanup, and repairs to reporting and taking photographs

Infrastructure Damages (*cont.*)	20,000 miles of electric and telephone cable	Used for repairs
		2,700 workers brought in by Bell Systems
	80% of power service down	For all New England
	30% of telephone service down	For all New England
	21,800 telephone poles replaced	
	79,000 additional phones accidentally disconnected after the storm	By homeowners and cleanup crews
Farm Losses	324,800 sugar maple trees down	20% of trees tapped for sugar syrup
	4 million bushels of apples picked by the storm	Roughly 400 million apples
	500,000 chickens lost	Average estimate
	1,675 livestock lost	
Forest Damage and Timber Losses	5 billion board feet of timber blown down	(12×12×1 inch board)
		Highest estimate (Lowest estimate 1.5 billion) Very roughly corresponds to 50 million trees
		Does not include trees not used for timber
	35% of New England covered in blown down timber	Approximately 15,000,000 acres
	1,256,278 board feet of salvaged timber	
	1,000 trails blocked in the White Mountains	10,000 acres of the White Mountains destroyed
Other Environmental Effects	17 inches—maximum total precipitation	Total from Hurricane plus preceding precipitation
		Average 11 inches over 10,000 square miles
	20 miles inland—vegetation killed by salt	Traces of salt reported as far away as 120 miles inland
	12 inlets created	Long Island

SOURCES: From various sources, some estimates, some actual numbers. Red Cross (1939), USGS (1940), Forest Service, New Hampshire Disaster Emergency Committee (2008) and RMS reports, Emanuel (2005), *National Geographic* magazine, and scientific articles by Jarvinen (2006), Tannehill (1938), Pierce (1939), and Brooks (1938).

The Hurricane also marked the beginning of a shift in the way in which the government dealt with natural disasters. Government agencies had an unprecedented role in the relief and rebuilding efforts and in the purchasing of goods (e.g., apples and timber) to avoid crashing prices caused by a disproportionate supply (see Chapters 6 through 8 for more information about the effects of the storm and the relief efforts that followed).[53]

The Great 1938 Hurricane clearly served as a catalyst for economic, governmental, environmental, and even social and cultural changes. From the shape of the coastline to the forest and mountain vistas, and everything in between, change was accelerated by the storm. The Hurricane did not distinguish among classes—the rich as well as the poor suffered greatly. A society that was very much set in its old ways was suddenly shaken to its core. Its physical structures (e.g., buildings, infrastructure) did not go back to what they were before and neither did its social structures, including well-defined roles and expectations driven by class.[54]

Change had been slow to come to the Weather Bureau during the early part of the 20th century, with the forecasting process having evolved little throughout the previous decades. In contrast, great advances in the understanding of how the atmosphere works and groundbreaking improvements in forecasting techniques led by Norwegian scientists had been happening in Europe since World War I.[55] The Bureau, housed under the Department of Agriculture and with a limited budget, was too overwhelmed with its ongoing agricultural and new (and fast growing) aviation duties, among others, to pay too much attention to the new developments or dedicate resources to atmospheric research—at least not in an organized, systematic way or by more than just a handful of individuals. Business as usual continued until the 1930s, when the agency was shaken off their stagnant ways. A special committee of the Science Advisory Board (SAB), created by President Franklin D. Roosevelt's administration shortly after he took office in 1933, finally urged the Weather Bureau to utilize the new methods emerging from Europe and, in the words of the Bureau Chief Willis Gregg, to "modernize its service in line with recent thought and practice elsewhere."[56]

The committee provided various recommendations toward that end, but perhaps the most important was the immediate adoption of a plan to implement the use of forecasting techniques based on the emerging field of what was then termed "airmass analysis." The knowledge of air masses and their behavior, both at the surface and in the upper atmosphere, is fundamental to properly determining how weather will change at any given location; and its application would eventually translate into significantly more accurate weather forecasts. Applying the recommendation, however, would require many additional resources in order to train both new and existing personnel and obtain more frequent and widespread observations. Updated methods and new instrumentation would also be needed to obtain appropriate upper-air data and more detailed surface data. Consequently, although immediately adopted, a suitable plan would take years to implement.

The Bureau did what it could to comply, using existing resources when possible. An examination was established for both new recruits and those seeking promotion to test them on knowledge relevant to airmass analysis. Additionally, efforts were put in motion to train a handful of forecasters who would then pass their knowledge on to others throughout the country. One of the first successful initiatives was collaborating with the U.S. Army and Navy to create an initial network of the much-needed upper-air observations through daily flights of instrumented airplanes (see Chapter 5 for more on airmass analysis, the SAB report, and the Bureau's efforts to implement the resulting recommendations). Full use of the new techniques (especially the use of upper-air observations) and implementation of the recommended changes in the day-to-day forecasting methodologies of the Bureau would have undoubtedly made a great difference in the actions of the forecasters in charge on the day of the Hurricane. Earlier in 1938 a program to send forecasters to study meteorology at academic institutions was announced.[57] An airmass analysis research division of the Weather Bureau, created in 1934 and housed in the Washington, D.C., Weather Bureau Headquarters Office, was also just now becoming more active.[58,59] Nevertheless, to many day-to-day forecasters, especially those with many years of experience, airmass analysis might have been seen as a bothersome novelty not worthy of their time—although to be fair, some of the most successful experienced forecasters of the time intuitively, but indirectly, did use airmass analysis principles in their forecasts. For example, Charles Mitchell, the forecaster in charge during the 1938 Hurricane and widely acknowledged as the best forecaster in the nation at the time, openly referred to those promoting the new techniques as "air messers."[60] However, he is recognized as having had an equivalent intuitive understanding of the behavior of the atmosphere, making his forecasts more accurate than was possible at the time, given the data available. This interplay between the old and the new ways played an interesting role in the events of September 21, 1938.

Only one Weather Bureau forecaster recognized the danger on the day of the Hurricane: a young man from the airmass analysis group who was reportedly filling in for a colleague from the forecasting section. His con-

Forecasters in business suits analyzed weather maps by hand on architect tables. Even though this picture inside the Washington, D.C., Weather Bureau Headquarters is dated 1926, little would have changed by the late 1930s. (NOAA Photo Library)

clusion was most likely guided by an analysis of the limited information on upper winds available. His assessment, however, did not match that of the forecaster in charge. Various versions exist of the story about how things played out between the two men on that day at the forecast office. Exactly what transpired, however, may never be known for sure. In the end, a hurricane warning was never issued, hurricane flags were never hoisted, and the nature of the impending storm was never accurately communicated to the public (see Chapter 5 for more information about the forecasting process and what happened at the Weather Bureau Office on that day).

The furious fallout from the failed forecast provided a needed push. Shortly after the storm, the Bureau was on the receiving end of angry letters and critical articles. There was also unrest within the inner ranks. It has been reported that a group of forecasters in the D.C. headquarters submitted a reorganization plan envisioning the forecasting office as a scientific laboratory and the need for training in the latest advances in the field, much in line with the advisory board recommendations of just a few years before.[61]

In any case, big changes were on the way. On September 14, exactly one week before the Hurricane, Weather Bureau Chief Gregg unexpectedly died of a heart attack while attending an aviation conference in Chicago.[62] Chief Gregg, in charge since 1934, had faced the difficult task of balancing the ongoing duties of the agency with the recommended changes, and strongly advocated for the higher budget needed to properly accomplish the improvements. One can only imagine how shaken the Bureau's Headquarters might have still been on the 21st, as the responsibility for forecasting the track of a looming hurricane was passed to their office. Only a week into his new position, it was Acting Chief Charles C. Clark who would have to answer to the public, the press, and the Secretary of Agriculture, and to defend the Bureau's actions and decisions.[63] By December, the new no-nonsense Chief Francis W. Reichelderfer (who would lead the agency through World War II and all the way to the early years of meteorological satellites) had been appointed.[64] Additionally, in 1940, the agency was moved from the Department of Agriculture to the Department of Commerce.[65] A new era of modernization was about to begin for the U.S. Weather Bureau. While the tragic nature of The Great Hurricane of 1938 was a product of the limitations of forecasting, technology, communications, and community preparedness that existed at the time (as will be discussed more thoroughly in subsequent chapters), in retrospect it proved to be a catalyst for remarkable improvements in all of these areas in the years following the storm.

2

THE TOOLS OF THE TRADE

Hurricane-observing tools and practices, and their role in 1938

To those who did not experience its impacts directly, the Great New England Hurricane of 1938 may have become lost in history, sandwiched as it was between the Great Depression and World War II. After the initial rush of reports, articles, and locally produced commemorative photo collections, decades went by before the storm's significant and lasting impacts both on New England and on the science and practices of meteorology started to be recognized. Before looking more closely at the Hurricane's meteorology and impacts, however, it is worth understanding the tools used to observe hurricanes in 1938, which undeniably played a significant role in the story of why the storm was such a surprise to the region.

Meteorology was a very different science in 1938 than it is today, with different practices and fewer tools at its disposal. Weather maps were painstakingly drawn by hand. Much less up-to-date information about the state of the atmosphere was available to forecasters, and forecasts took much longer to develop and reach the public. The lack of current data meant that forecasts were by necessity heavily influenced by the intuition and experience of the one forecaster in charge at any specific time. It should not be surprising, then, that the ability to detect, monitor, and forecast hurricanes (as well as any other type of weather phenomenon) was much more limited (see Chapter 5

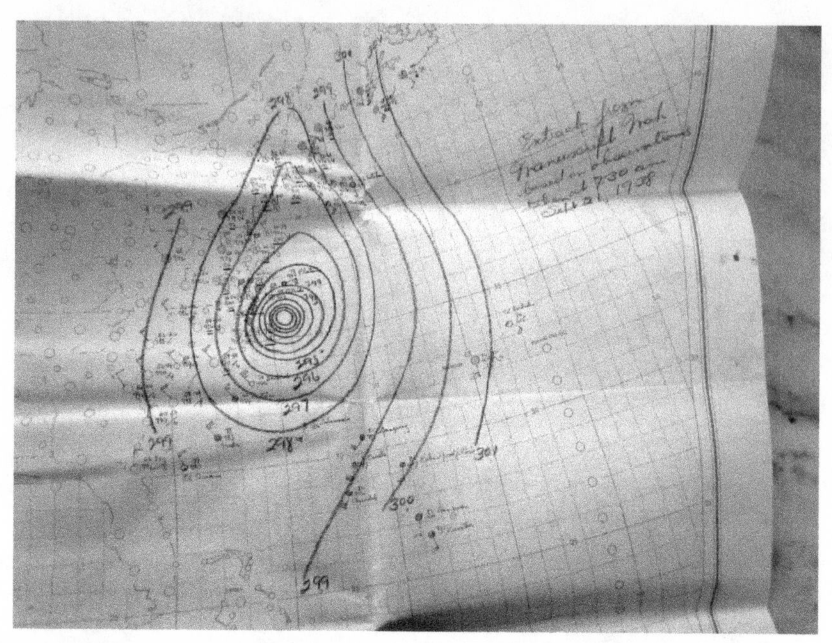

Surface weather charts of the Hurricane were put together by using available data from ships and land stations. A polished hand-drawn version was included with a report from the Weather Bureau to the Secretary of Agriculture two weeks after the storm. The contours are lines of constant pressure in inches of mercury, drawn every 0.1 inches. The handwritten caption reads, "Extract from manuscript map based on observations taken at 7:30 AM Sept 21, 1938." Notice the lack of data mostly to the east and northeast of the storm. The original hand-drawn map (approximately 3 feet high by 2 feet wide) is housed in the Department of Agriculture textual records at the National Archives and Records Administration in College Park, Maryland.

for more on the history of hurricane forecasting and the role of the Weather Bureau on the events of September 21, 1938).

Hurricane Monitoring and Warning

The need to adequately warn of threatening hurricanes was recognized as early as the late 1800s, when, with very little observational data available, simply determining their location was a challenge. This, of course, was especially dangerous to ships at sea. In preparation for the Spanish–American War, and at the urging of Weather Bureau Chief Willis Moore (who feared for the safety of the American Navy), Congress approved the addition of observation stations in the Caribbean in 1898. This was just two years after a deadly hurricane had swept from Florida to Pennsylvania, killing more than 100 people in the process. Even President William McKinley is reported

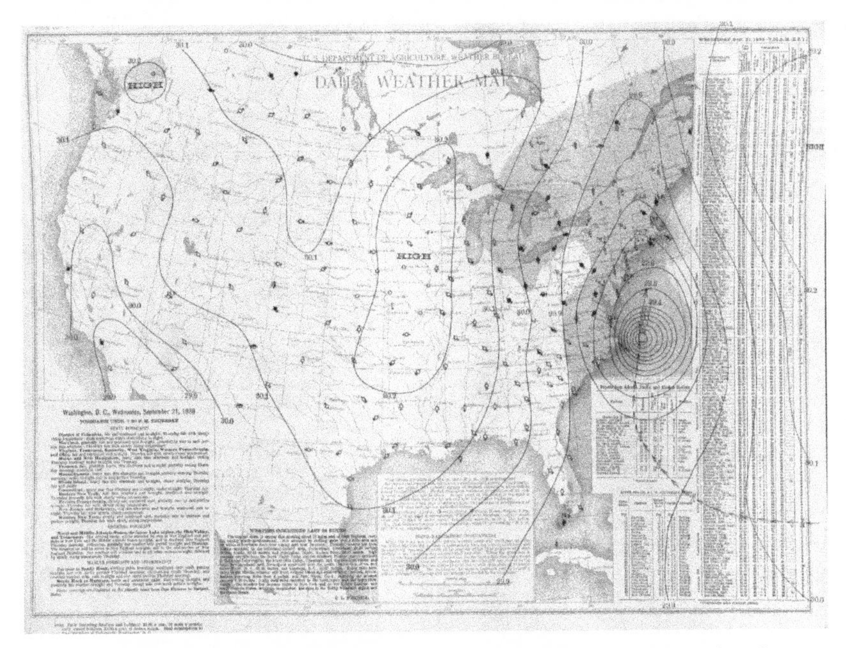

Map summaries of the weather for each day have been published since 1871 (when they were called War Department Weather Maps). The U.S. Department of Agriculture Weather Bureau Daily Weather Map for September 21, 1938, depicts conditions in the early morning of the day of the Hurricane's landfall. (NOAA Central Library)

to have said that he feared a hurricane more than he feared the Spanish navy.[1] Early on, all warnings were issued from the Washington, D.C. Office, a process that would prove cumbersome and inadequate throughout the early 1900s. In response, a more organized and decentralized system was established in the mid-1930s (see Chapter 4 for more about the history of hurricane warnings and the Hurricane Warning Service used by the Weather Bureau in 1938), but it was not until the formation of the National Hurricane Center (NHC) two decades later that modern-day hurricane warning and forecasting started to take form.

Today, the NHC is in charge of hurricane and tropical storm watches and warnings along the U.S. coast as well as advisories, forecasts, and discussions for each storm.[2] The Center was established in the 1950s, when the primary hurricane forecast office (which had been moved in 1943 from Jacksonville to the Weather Bureau Office in Miami) and the newly created National Hurricane Research Project were brought together into one entity devoted to forecasting and research.[3] The agency's creation was ultimately a response to the 1954 hurricane season, which had just broken the record as the costliest

TROPICAL CYCLONE ALERTS

Hurricane and tropical storm watches and warnings are announcements issued according to the possible sustained (one-minute average) winds expected for the specified area and due to a tropical, subtropical, or post-tropical cyclone. Because outside preparedness activities become difficult once winds reach tropical storm force, these alerts are issued before the anticipated onset of tropical storm–force winds, 48 hours in advance for watches and 36 hours for warnings.* The time and practice specifics of issuing these alerts are revised from time to time to improve clarity and safety.†

- Tropical Storm Watch—tropical storm conditions (39 to 73 mph) possible
- Hurricane Watch—hurricane conditions (at least 74 mph) possible
- Tropical Storm Warning—tropical storm conditions (39 to 73 mph) expected
- Hurricane Warning—hurricane conditions (at least 74 mph) expected

Once in place, a hurricane warning can remain in effect when dangerously high water or a combination of dangerously high water and waves continues, even if winds are no longer of hurricane force.

---

* For more information and related links, see www.nhc.noaa.gov/prepare/wwa.php.

† The Tropical Cyclone Definitions (National Weather Service Instruction 10-604) contains the most up-to-date information (www.nws.noaa.gov/directives).

to date.[4] After Hurricanes Edna, Carol, and Hazel swept through the East Coast in a matter of just two months, Congress finally recognized that scientific understanding of hurricanes and a dedicated agency were needed if more accurate forecasts were ever going to be possible.

The Hurricane Warning Service had only been in place for three years when the Hurricane came. Starting the moment when the storm that would become the Great New England Hurricane of 1938 was detected at sea, the Jacksonville, Florida, Bureau Office was the first in charge of issuing advisories and warnings. Then, as the storm moved past Cape Hatteras during the early morning before the Hurricane's landfall, the responsibility was passed to the D.C. Central Office. (See Chapter 4 for a map illustrating monitoring and forecasting responsibilities of each Bureau Office and Chapter 5 for more on the hurricane's forecasting by the D.C. Office.)

Ships, Planes, and Hurricanes

Today it is unthinkable that a powerful hurricane threatening a major center of population could go undetected or ignored. Satellite images allow forecasters to observe a hurricane as it develops from innocuous tropical disturbances, and broadcasters provide blow-by-blow descriptions of a hurricane's changing appearance, intensity, and location. These observations are possible thanks to meteorological and environmental satellites managed by the National Oceanic and Atmospheric Administration (NOAA) and the National Aeronautics and Space Administration (NASA). These remarkable tools constantly survey the atmosphere (as well as other elements of the earth system) and can be used not only to visually detect hurricanes but also to measure their temperature, moisture, precipitation, and even the roughness of the ocean's surface during their passage. These data provide invaluable help with forecasting storm evolution, but satellites, of course, were not available in 1938. It would take a few more decades. After a handful of failed attempts, an image of the atmosphere was finally recorded from space on April 1, 1960, and it was not until the late 1970s that reliable, widespread, and continuous meteorological satellite coverage was in place, allowing us to instantly see most tropical cyclones around the globe.[5]

The famous hurricane hunters—NOAA and Air Force personnel who fly straight into the center of the storms—also currently provide essential data to tropical meteorologists. Their observations are used to improve track and intensity forecasts as well as to research various aspects of the nature of hurricanes. During the presatellite years, this provided the most accurate way to locate hurricanes (and it is still the most accurate way to measure their intensity). The appropriate title of "hunters" was earned during the early years when, armed with only approximate position information determined from wind and pressure observations provided by ships, plane crews set out to hunt for the core of a hurricane by relying solely on their own visual observations. In 1938, however, no airplanes would approach the Great New England Hurricane; there would be no way to correct the wildly erroneous location and intensity estimates. During the 1930s, visual clues were still essential to flight, and navigating through bad weather, which would obstruct a pilot's view, was not an option. It was not until 1940 that development of methods for instrument-assisted flying began in support of the war efforts.

Airplane surveillance of weather systems was routinely done by the U.S. Army Air Force starting in 1942, but flying intentionally into a hurricane would most likely have still been considered an insane exercise at that time.

Aircraft reconnaissance of hurricanes might never have been attempted if it was not for an unauthorized flight resulting from a friendly challenge. Army Air Corps Colonel Joseph B. Duckworth had left his job as an airline pilot to join war preparations and was involved in the establishment of a flying school in Bryan, Texas, where he specialized in teaching military pilots how to fly during bad weather by using instrumentation rather than visual cues. On July 27, 1943, a hurricane lurking off the island of Galveston prompted an order to move all of the school's AT-6 aircrafts to safety. A group of British students (unfamiliar with both hurricanes and the planes' capabilities) taunted their instructors about the ability of the small airplanes to withstand bad weather. Col. Duckworth bet the naysayers that he could fly into the hurricane and back with an AT-6 without incident—and he did, twice on that day! The same hurricane went on to seriously damage refineries in Texas, significantly affecting fuel production—an especially serious problem during wartime. Just a year later, in recognition of the threat that hurricanes posed to Navy vessels, cargo ships, and other services important to wartime production, a reconnaissance program (which continues to this day) was officially established for the Atlantic Ocean.[6] Today, the responsibility of aircraft reconnaissance of hurricanes in the Atlantic Ocean, Caribbean Sea, and Gulf of Mexico is shared by the Air Force Reserve's 53rd Weather Reconnaissance Squadron and NOAA's Aircraft Operation Center.[7] Specially equipped propeller planes (such as NOAA's WP-3D Orion and the Air Force's WC-130J Hercules) are used to measure wind, pressure, temperature, and dew point, as well as to drop instrument sondes to measure these same quantities at lower heights, as the airplanes fly from the outskirts to the center of hurricanes and back.[8] Their missions include both routine monitoring of threatening storms and intensive research field campaigns. Even in the age of sophisticated meteorological satellites, airplane reconnaissance still provides the most accurate hurricane observations possible, essential to better modeling and forecasting.

In 1938 and throughout the preceding decades, the Weather Bureau relied on ships at sea (those in the wrong place at the wrong time) to gain information about the position and strength of hurricanes, or even about their very existence. Therefore, the amount and quality of information being received by forecasters was wildly variable. Over the years, the agency developed relationships with private companies and other countries, whose ships would take regular observations (twice a day) as they traversed the portion of their routes within known "hurricane districts" and mail them in once they came back to port. Once radio communications became commonplace (at the turn of the 20th century), it was not long before most vessels were properly equipped for what was then known as "wireless telegraphy" (which consisted of transmitting a telegraph-type message via radio waves rather than through cables). By 1938, the Bureau was receiving both mailed-in and radio observations from ships with which it had agreements. A "calling program" had also been established, by which forecasters could call by radio any ship known or suspected to be in the vicinity of a storm in order to gain additional information.[9]

The amount of information forecasters learned about the Great Hurricane of 1938 (and other presatellite storms) from just a handful of weather measurements from ships caught in the path of the storm is remarkable. The most important observations obtained were atmospheric pressure (usually reported in inches of mercury) and wind direction and intensity. These seemingly simple observations could be used to estimate information such as a storm's position, maximum winds, wind distribution, and minimum pressure in the center of the hurricane (inversely related to the intensity of the hurricane). But observations could not be methodically distributed in a useful pattern; they were taken wherever ships happened to be. Some came from ships smack in the center of a storm and some from ships well outside its periphery. Further, different types of instruments were used, preventing assurance of quality or consistency. Assumptions as to what individual observations indicated about actual storm characteristics were made by mathematical equations known as pressure-wind relationships.[10] Depending on the number of observations available, the result of the calculations was usually a good, but not perfect, approximation of the position and intensity of the storm. Data from as many ships as possible were obtained and aggregated in an attempt to determine the storm's position (using triangulation of wind directions, and assuming a 20-degree inward angle for winds outside the center of the storm) and strength (using pressure measurements from the center of the storm).[11]

The more ships reporting, the more accurate the estimates of position and intensity. The problem with this method is that once the presence of a hurricane was known, ships would steer away from it, leaving the area in which observations were most needed; these data-gathering arrangements were voluntary and the first priority would have been the safety of passengers and cargo. Based on the information obtained from ships at sea, then, many advisories for the 1938 Hurricane were issued from the Jacksonville Office as the storm headed toward Florida (see Chapter 4 for more information about the advisories issued by the Jacksonville Office). By the time the storm raced northward, there were barely any reporting ships left out to sea.[12] Therefore, the forecasters' estimates of the position, movement, and intensity of the storm were inaccurate enough to make a difference in the interpretation of the imminent threat (see Chapter 5 for more information about the accuracy of the estimates of the storm position and movement).[13]

Observing the Weather Aloft

It is not enough to know where a hurricane is; somehow forecasters must determine where it is going next. As will be discussed later, hurricanes are steered by the larger-scale environment in which they are moving, and the more easily obtained surface wind measurements do not provide enough information to forecast their movement accurately.

Kites and manned hot air balloons carrying thermometers and other instruments had been used since the 1700s to take atmospheric measurements above the surface, but mostly these were novelty experiments. Similarly, shortly after the establishment of the Weather Bureau in the late 1800s, specially designed kites and small balloons (called pilot balloons) were used to measure some weather conditions at modest altitudes.[14] In 1929 Robert Goddard (for whom the NASA Space Flight Center in Maryland is named) launched a thermometer, barometer, and a camera on a rocket.[15] The camera was set to record the instrument readings the moment the parachute deployed, signaling the end of the ascending flight (this experiment also showed the feasibility of high-altitude photography, which later led to the development of satellites). In the 1930s, kite stations were phased out and replaced by airplane stations, where observations were taken during short flights to higher altitudes than those possible with tethered kites, though still within the troposphere (the layer of the atmosphere closest to the surface).

At the same time, larger hydrogen and helium weather balloons with attached instruments had been successfully developed to provide a better way to obtain data up to much higher altitudes (about 50,000 feet, within the lower stratosphere, the very stable layer above the troposphere, where temperature on average increases with height). The measuring instruments created what was then known as a meteograph, a line chart showing weather conditions as the balloon ascended. The problem is that to examine the measurements, the meteograph had to be retrieved after the balloon burst and fell to the ground. Retrieval required time and luck, and often proved

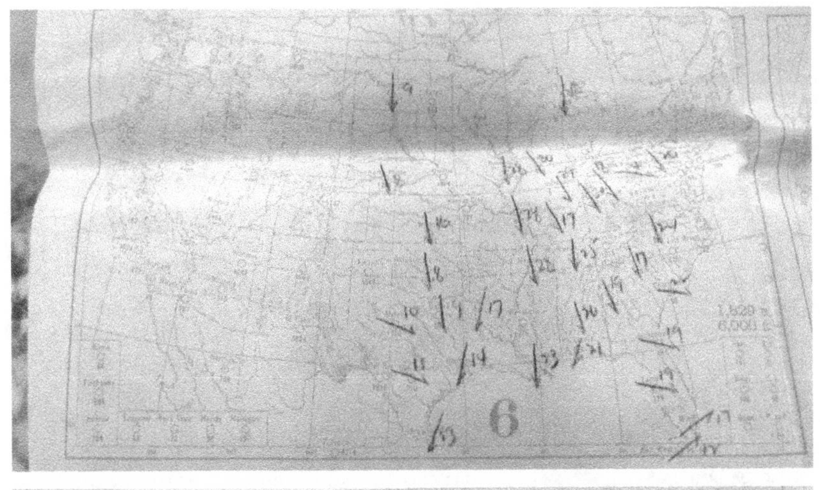

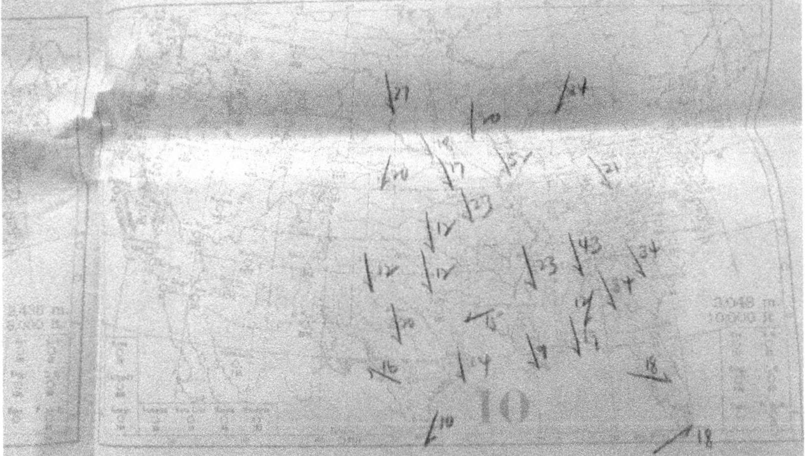

Upper-air wind vectors based on observations at 6,000 feet (top) and 10,000 feet (bottom) show northerly winds in the central to eastern United States on September 21, 1938. The original hand-drawn maps are available in the Department of Agriculture textual records at the National Archives and Records Administration in College Park, Maryland.

impossible, depending on where the device landed. The solution came in the form of what was first called the "radiometeograph." As the name suggests, this was an instrument with radio transmission capabilities, allowing for the instantaneous relay of information to the ground. Because no actual graph is produced in flight, the name was quickly deemed unsuitable and changed to "radiosonde."[16] The first weather balloon radiosonde in the United States was actually launched in New England, from East Boston, Massachusetts, the year before the Great New England Hurricane.[17]

TABLE 2.1. Main Hurricane Observation Tools: Then vs. Now

Observation Tools	1938	Today	History of Availability
Ships	x		Limited availability today, but not relied upon
Upper-air	x	x	Limited in 1938 Airplane and balloon obs in 1938, only balloons now Balloon network mostly in Central Plains in 1938
Satellite		x	Full global coverage in 1979 Various types of detectors used to measure different aspects of hurricanes. Various types of satellite sensors and observations have been added throughout the years.
Radar		x	First used to observe hurricanes in the 1940s
Hurricane hunters		x	Hurricane aircraft reconnaissance program established in 1944
High-altitude jets		x	Started in 1997
Unmanned aircraft		x	Aerosonde observations started in 1998—currently in experimental stage

A more widespread and coordinated upper-air observations network, which continued to grow throughout the years, was first formally put in place just a year after the Hurricane hit. If available in 1938, plentiful upper-air observations would have indicated the potential for the Hurricane to be steered northward. Instead, most of the limited upper-air information (which came almost exclusively from airplane stations) was not available for analysis until after the storm had passed. The lone forecaster Charles H. Pierce, who is known to have predicted the Hurricane would endanger the northeast United States instead of turning toward the North Atlantic as expected, used a very small amount of information, together with knowledge of new analysis techniques, to reach his conclusion. Unfortunately, precedence and extensive experience with previous storm tracks, which weighed much more heavily in the eyes of the senior forecasters on that day, did not indicate the possibility of such a track, and in any case it was most likely already too late to disseminate adequate warnings to the threatened area. (See Chapter 5 for more information on the events that transpired at the Washington, D.C., Weather Bureau Office on September 21, 1938.)

As shown in Table 2.1, with tools such as satellite and radar imagery, high-altitude jet observations, and others, the monitoring of hurricanes (and the use of the obtained information in forecasting and warning) has undergone countless improvements, especially in the era of computers. Advances have been vast and extremely beneficial to those in the potential track of powerful hurricanes. In 1938, however, New Yorkers and New Englanders would not benefit from most of these tools.

Reanalyzing Historical Storm Data

The most up-to-date track and intensity information about the 1938 Hurricane, contained in the NHC's HURDAT/2, is the result of the reexamination of all of the available data for this storm done as part of a much larger project. But why even look back at storms long past? The reasons to revisit historical hurricane data are many, from creating the most accurate picture of a specific storm or season, to performing climatological studies of the geographical initiation or landfall of tropical cyclones, to determining cycles and trends in frequency and intensity. HURDAT/2 can and has been used as a research tool for many of these and other applications.

In order for a database to be useful, especially if one wants to use information for scientific research, it needs to be constructed in a consistent manner. Unfortunately, the tools and methods used to observe tropical cyclones and estimate their position and intensity have changed significantly throughout the years. Consequently, the historical database is at best uneven and incomplete, especially for the presatellite period. Additionally, new information can come to light well after the end of a hurricane season, sometimes many years afterwards, through the examination of records (such as ship journals and other types of historical documents). The Atlantic Hurricane Reanalysis Project (briefly introduced in Chapter 1) is immensely important to making records consistent. This ambitious undertaking, initiated in 2001 under the leadership of Chris Landsea (then Research Meteorologist at the Hurricane Research Division of NOAA's Atlantic Oceanographic and Meteorological Laboratory and currently Science Operations Officer at the NHC), is a collaboration among hurricane scientists from all over the country.[18] The overall methodology consists of subjecting old data to state-of-the-art and uniform analysis techniques, including updated and regionally specific pressure–wind relationships. Any new information that has surfaced is folded in. The process results in the most complete and consistent database possible. Once the information for a specific storm, season, or decade is individually revised, it is then submitted to a special NHC group, the Best Track Change Committee, which reviews submitted changes, provides suggestions, and ultimately approves or rejects deletions, additions, and alterations to be incorporated into HURDAT/2. Fortunately, reanalysis results for the 1938 hurricane season were finalized and approved in December 2012, and so the most updated information has been included here.[19]

To fully appreciate the story of the great storm that simultaneously devastated New England and nudged meteorological science forward, we must return to early September 1938, when a seemingly harmless disturbance was detected over Africa. We will then follow the disturbance and eventual hurricane as it traveled over the ocean as well as the unfolding events that led to the unexpected landfall and passage through Long Island and New England. In the process we will learn about the meteorology of the storm's behavior. Finally, we will examine what was left behind by the storm and what it all means for potential future hurricanes in the region.

PART II
LIFE CYCLE

3
BIRTH OF THE STORM

Evolution from a harmless African disturbance to a full-fledged Atlantic hurricane

Like the vast majority of the most intense Atlantic Ocean hurricanes, the Great New England Hurricane owed its origin to a type of tropical weather disturbance known as an African easterly wave.[1] Every year, as few as 50 to as many as 75 such disturbances form from May to November.[2] A few of them at some point provide a perfect development environment for many of the region's hurricanes.

An African Wave

In the deep tropics, weather systems travel from east to west within the same trade winds that brought Christopher Columbus to the New World more than 500 years ago. The dozens of tropical disturbances that originate over northern Africa every year are thus also steered westward. Historically, humans have named the wind by using the direction from which it is blowing. Likewise, in meteorology—a science that must communicate effectively with the public—wind direction is defined in the same manner: a north or northerly wind comes from the north and a northeasterly wind comes from the northeast.[3] The trades are, therefore, easterly winds. The practice further applies to the direction of motion of a weather system. This is why the hurricane-spawning tropical disturbances that develop over Africa and travel westward over the Atlantic Ocean are commonly known as easterly waves.[4]

In a way, easterly waves can be considered the tropical counterparts to the extratropical cyclones introduced in Chapter 1. Their size, for example, is comparable, roughly one to two thousand miles wide. Even more important, however, is that their initial formation and energy source are rooted in temperature contrasts between the extremely hot Sahara Desert and the cooler grasslands and ocean to the south of the desert (in much the same way that the formation and energetics of extratropical cyclones are associated with temperature contrasts across fronts).[5] On a map, they appear as wavy perturbations in the easterly trade wind flow. On satellite images, they sometimes resemble an inverted "V" as clouds wrap over the axis of the north–south oriented and inverted trough, straddling the clear (west side) and cloudy (east side) regions. (In meteorology, a trough is defined as an elongated area of low pressure, which, as the name suggests, normally looks like a valley or dip in the contours on a weather map. Within the deep tropics, however, troughs manifest upside down, taking the inverted-V shape.)[6]

The passage of a series of these waves over a location produces an alternating cycle of weather conditions over a period of three to five days. In its path, a tropical island would experience one or two sunny days with slightly lower humidity while the barometric pressure slightly decreased, followed by one or two stormy—or at least rainy and more humid—days while the pressure slightly increased.[7] The weather pattern would then be followed by the same or a similar sequence as the next wave came through. This basic model of the structure and weather conditions associated with an easterly wave was first proposed in the 1940s by renowned tropical meteorologist Herbert Riehl, who named them "waves in the easterlies."[8,9] He based his description on disturbances observed over the Caribbean Sea, mostly using upper-air observations from San Juan, Puerto Rico. His simple model does not apply perfectly to all easterly disturbances now considered easterly waves, but it provides a very good conceptual understanding of these systems that are so often involved in hurricane formation.

Most African easterly waves harmlessly cross the Atlantic Ocean, bringing a good portion of the normal yearly rainfall to the islands of the Caribbean. However, a handful of them end up providing the preexisting low pressure disturbance needed to produce a tropical cyclone. With more than 50 crossing the Atlantic every year, it is fortunate that only a relatively small percentage are associated with tropical cyclone formation in the area (less than 20 percent for all tropical cyclones and only about 5 percent for storms that eventually intensify into hurricanes).[10] On the flip side, the percentage of Atlantic hurricanes originating from easterly waves is quite high. There

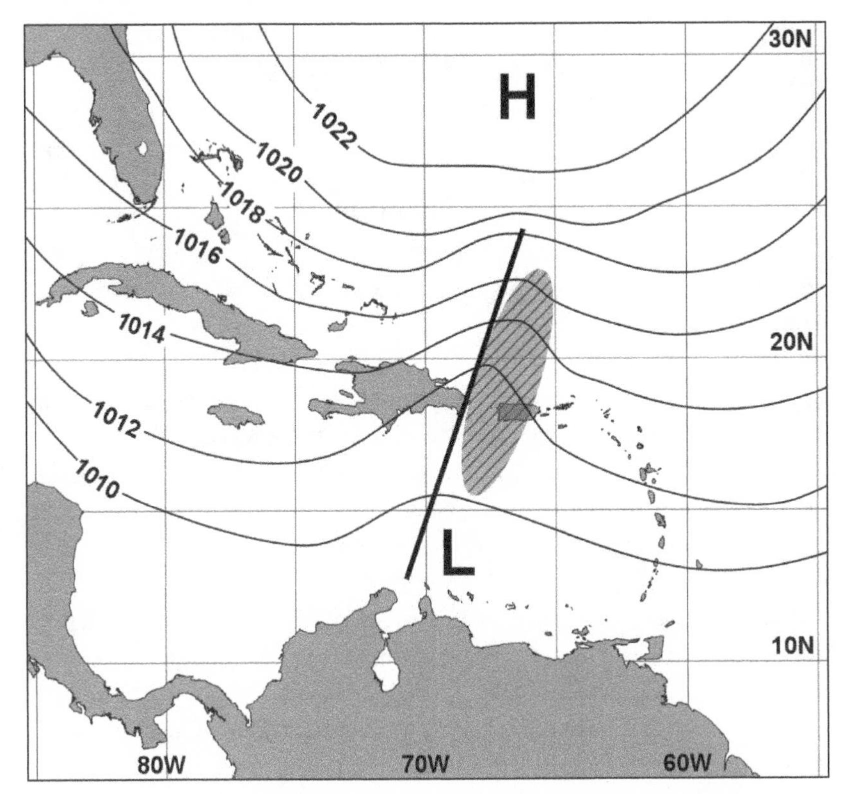

The classic easterly wave model shows a kink in the contour lines of pressure (isobars), an inverted trough (marked by the solid line), and a region of cloudiness and rain (shaded/hatched) to the east of the trough axis. (Avilés [2004], adapted from Riehl [1954])

is significant variation from year to year, but the average is approximately 65 percent.[11] Tropical cyclone formation is a complicated phenomenon, but what the African waves can ultimately provide is a favorable low pressure and thunderstorm environment in which the mechanisms needed to form the rotating, intensifying storm can occur. The formation of the Great New England Hurricane of 1938 was most definitely associated with an easterly wave, and all signs point to this wave forming and crossing northern Africa at the beginning of September.

The availability of weather observations over the African continent is rather scarce in comparison with North America and Europe. This is especially true over the Sahara Desert and its southern fringes, the same region where African waves form and start their westward travel. In 1938 easterly waves had not yet been identified as such, but there was at least one location where the observed weather corresponded to the passage of one such disturbance, a wave that would later be connected to the Great Hurricane.[12] The observations were made at the Bilma Oasis in Niger, in the south-central part of the Sahara Desert. Niger was then a French colony, and a French weather observer stationed there recorded signs of a disturbance on September 4.[13] It probably was as innocuous as a slight shift in wind direction, maybe accompanied by rain or even thunderstorms (possibly on the intense side, but not much different from many other disturbances regularly passing over the same region during that time of year). The description of what happened next comes from our knowledge of the typical characteristics and behavior of easterly waves.

As the disturbance inconspicuously moved westward in the easterly flow, it intensified and became deeper (meaning that it reached higher into the troposphere), accompanied by increasing thunderstorm activity as it neared the western coast of Africa and crossed onto the Atlantic. Until then, its energy had come largely from temperature contrasts over Africa, but now, fully over the ocean, it obtained most of its energy from the condensation of water vapor readily available within the tropical maritime atmosphere.[14] Most waves immediately weaken after exiting the African continent (later regaining some of their intensity during their westward travel over the ocean), but in this case its thunderstorms persisted and quickly developed a counterclockwise circulation, thus becoming a tropical cyclone.

Tropical Cyclogenesis

The formation of a tropical cyclone, known to meteorologists as tropical cyclogenesis, is a complicated process with nuances that are still not fully understood. The broad strokes of such a process, however, can be described in relatively simple terms.

To understand how a group of thunderstorms within a tropical disturbance begins to rotate requires knowledge of the Coriolis effect. As air moves through the atmosphere and across large distances, Earth rotates underneath it, which, from our perspective on one point of the rotating Earth, results in the apparent deflection of air. This deflection is known as the Coriolis effect—also referred to as Coriolis force or acceleration. Because of Earth's relative direction of rotation around its terrestrial axis, this deviation from the air's original direction of motion is toward the right in the Northern Hemisphere and toward the left in the Southern Hemisphere.[15,16] How this effect causes a disturbance to rotate is not as immediately obvious as one might think. Air flowing through a low pressure disturbance—such as an easterly wave—converges or piles together at the lower levels of the atmosphere. The accumulated air is then forced to rise, which results in cloud formation (on the east side of the shallow trough in the case of the classic easterly wave model).[17] But as the air converges, it is also being deflected by the Coriolis effect. Air coming from the north is deflected westward; air coming from the south is deflected eastward; air coming from the east is deflected northward, and so forth, initiating a counterclockwise flow. However, this incipient rotation is not enough to produce the self-sustaining, powerful, and stable vortex that is a hurricane. Additionally, not only Coriolis but other atmospheric forces work together to produce the resulting air flow around the storm. There is much more, then, to the story of tropical cyclogenesis as will be discussed in Chapter 4. Nevertheless, for now, we can say the disturbance that would become the Great New England Hurricane exited the African continent as a strong tropical wave and quickly started to rotate and intensify.

A Cape Verdean Hurricane

The Republic of Cape Verde is a group of islands roughly three to four hundred miles off the western coast of Africa.[18] The name might suggest a green luscious landscape, but its rocky and gravelly terrain is quite dry, as the islands receive infrequent rainfall, totaling a desertlike 10 inches per year at the capital of Praia. Extended droughts are also quite common. On average, most of the precipitation falls from August to October, with September experiencing the highest amounts.[19] Portuguese explorers who came upon the islands in the 1450s and 1460s named them Cabo Verde (Green Cape) in reference to the westernmost point of West Africa with which the islands align, a peninsula then known as Cabo Verde itself. Today the peninsula is called Cap Vert and is home to the Senegalese capital, Dakar. (The modern name Cap Vert also means green cape, and dates back to the French colonial era of the region.)[20] Both the Cape Verde islands and the Cap Vert peninsula lie directly on or close to the path of easterly waves moving off Africa, and the date when a wave known to later initiate a tropical storm or hurricane passes through Dakar often serves as a standard reference point in post-mortem Atlantic Tropical Cyclone Reports prepared by the National Hurricane Center (NHC).[21]

The first appearance of the 1938 storm within HURDAT2 is as a tropical depression just off the African coast and east–southeast of the Cape Verde islands during the morning hours of September 9. As shown in Table 3.1, the storm quickly intensified into a tropical storm and then into a hurricane of increasing strength as it crossed the Atlantic Ocean. Such classifications, however, were not yet in place in 1938, not to mention that it would have been

TABLE 3.1. Great New England Hurricane Timeline

Date	Intensity kt (mph)[a]	Category[b]	Qualifier	Comment
Sept 4	-	EW disturbance	Disturbance detected at Bilma Oasis in south/central Sahara	Would not have been recognized as an "easterly wave," since it was not until 1940 when they were identified as such but the connection between the Hurricane and its original disturbance was recognized soon afterward.
Sept 8–9	-	EW/TD disturbance	Estimated emergence from Africa and fast development as a depression	Passage near Dakar, Senegal, used today as a reference point in tropical cyclone reports
Sept 10	35 (40)	TS	Storm near the Cape Verde Islands	Rapid development from disturbance to tropical depression to tropical storm after emergence from Africa
Sept 15	65 (75)	HR1	Central Atlantic	Estimated a posteriori from analyzing available ship observations
Sept 16	85 (95)	HR2	Western, Atlantic/off NE Caribbean	
Sept 16	-	-	Reported by S.S. *Alegrete*	Earliest report made of a hurricane obtained by the Weather Bureau Report not live, but while Hurricane still at sea
Sept 16	100 (115)	HR3	Major hurricane	Rapid intensification just as its existence was being discovered
Sept 17	-	-	Reported by M.S. *Socrates*	First real-time observation of the Hurricane radioed to the Weather Bureau
Sept 17	-	-	1st Jacksonville advisory	Total of 15 advisories from Jacksonville, from detection (when it looked like it was heading to FL) to morning of the 21st when the storm was south of Cape Hatteras
Sept 17	115 (130)	HR4	-	Afternoon of the 17th
Sept 17–20	-	-	Prehurricane rain event in New England	Left ground saturated and many rivers near or above flood stage
Sept 19–20	140 (160)	HR5	Estimated maximum intensity of the storm	East of the Bahamas, north of the Caribbean

Date	Intensity kt (mph)[a]	Category[b]	Qualifier	Comment
Sept 21	-	-	1st Washington D.C. advisory	First storm advisory in the morning (storm warnings ordered north of Cape Hatteras previous day) Seven advisories from D.C. Warnings up to New Jersey upgraded to “whole gale” at 11:30 A.M. No hurricane warnings issued.
Sept 21	-	-	Last Washington D.C. advisory	Last advisory at 2 P.M. mentions “tropical storm” will pass over Long Island and Connecticut late afternoon. Northeast storm warnings had been issued in the morning up to Eastport, ME.
Sept 21	105 (120)	HR3	Long Island landfall	2:45 PM (EST)[c]—Near Bellport, Long Island
Sept 21	100 (115)	HR3	Southern New England landfall	3:40 PM (EST)[c]—Near New Haven, Connecticut
Sept 21	60 (70)	TS-intensity (post-tropical)[d]	Extratropical transition (see Chapter 7)	Exact moment of full transition to extratropical cyclone not known. Most likely transitioning during landfall and fully transitioned while over northern New England.
Sept 23	30 (35)	TD-intensity (post-tropical)	Dissipation over Canada	Fast weakening from intense hurricane strength to tropical depression strength once center fully over land
Sept 23	-	-	End of flood event	Prehurricane plus hurricane precipitation caused maximum river levels Sept 21–22

NOTE: Information gathered from various sources, including the official NHC hurricane records (HURDAT2), updated for the 1938 hurricane season in December 2012.

[a] Hurricane wind speeds normally reported in knots (kt), approximate mph conversion included here for clarity.

[b] EW—easterly wave, TD—tropical depression, TS—tropical storm, HR—hurricane (the number signifies the Saffir–Simpson category)

[c] At the time of the Hurricane, RI, CT, NH, and NYC used EDT, but VT and DC used EST, which results in 1-hr inconsistencies encountered in reports of the storm's landfall.

[d] “Post-tropical” terminology was adopted in 2010 to refer to previous tropical cyclones that no longer possess sufficient characteristics to be considered a tropical cyclone (more about this in Chapter 7).

impossible to observe the sequence of events or to verify the occurrence of any specific classification criteria. However, it is safe to say that the easterly wave observed at the Bilma Oasis, which moved off Africa as a tropical disturbance just a handful of days later, very soon developed a rotation and intensified into the Hurricane that was later encountered by ships at sea in the western Atlantic.

Today, Atlantic tropical cyclones that form in the vicinity of the Cape Verde islands and then move across the Atlantic Ocean, intensifying into hurricanes that threaten the Caribbean islands or the United States, are informally known as Cape Verde hurricanes.[22] The Great New England Hurricane of 1938 was without a doubt a Cape Verde hurricane.

An Invisible Storm Found

It is important to understand that even though the observations at the Bilma Oasis and the Cape Verde islands were made, they were not connected to the New England Hurricane until later analysis became available. The same is true for all ship observations indicating a storm during the next few days. From these, using the methods described in Chapter 2, meteorologists have determined a posteriori that the storm most likely intensified into a hurricane during the evening of September 14 in the central Atlantic Ocean (when located near 18°N, 44°W).[23]

The earliest report of a hurricane came from a Brazilian ship, the S.S. *Alegrete*, a giant cargo steamer that was being used to train the Brazilian Merchant Marine and that conducted regular routes through the northeast Caribbean. The *Alegrete* has a dramatic history in its own right: it was seized by Brazil during World War I after it torpedoed too many Brazilian ships, and it ironically met its demise when hit by a torpedo in 1942. Brazil was neutral during World War II, but the ship's flag was not visible enough to prevent it from becoming a target.[24]

According to the original storm report appearing in the September 1938 issue of the *Monthly Weather Review* (MWR):

> At about 9:30 P.M., ship's time, on September 16, the Brazilian S.S. *Alegrete* was near the center [of the storm] in approximately 21°12' N., 52°46' W., barometer 28.31 (uncorrected), wind force 12, shifting from east-northeast to east-southeast.[25]

This position near 21°N, 53°W locates the storm a few hundred miles northeast of the Caribbean. The "barometer 28.31" observation refers to

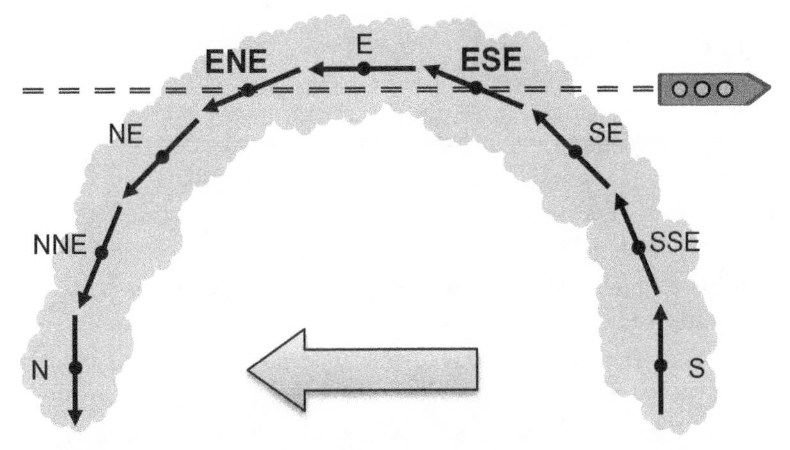

The Brazilian cargo steamship *Alegrete* reported wind direction shifting from east-northeast to east-southeast. This suggests that it might have skirted the northern portion of the hurricane's eyewall. (The cloudy semicircle represents the northern half of the eyewall. The small arrows show the expected wind direction at each eyewall location, identified by the compass direction that a ship would report at that position. The large arrow represents the general direction of motion of the Hurricane. If the ship was traveling outside the eyewall, within the spiral bands of the storm, the wind arrows would cross inward at a 20°angle instead of being parallel to the circle.)

inches of mercury (the height of a column of mercury pushed upward by the weight of the overlying air in a classic barometer, a commonly used atmospheric pressure unit). In today's standard units preferred by meteorologists, the reported pressure would correspond to approximately 955 hectopascals, a likely sign of an intense hurricane.[26,27] The pressure is also identified as "uncorrected," meaning that it has not been adjusted for altitude as is normally done; this would not be as big a problem in the case of an observation taken at sea, in other words, at no altitude. The shift in wind direction from east-northeast to east-southeast indicates that the ship, assumed to be close enough to the center of the storm, might have moved from the northwest to the northeast quadrant of the eyewall or its vicinity, where these wind directions would each be expected.[28]

The "wind force" value refers to the Beaufort Scale, developed in the early 1800s by British Admiral Francis Beaufort, a method very commonly used during the 19th and early 20th centuries to estimate the strength of the wind.[29] The wind force "12" reported by the *Alegrete* is the highest possible and signifies a sea that was completely white with foam and spray, with huge waves and greatly reduced visibility. Only winds blowing with the strength of a hurricane could cause such effects on the surface of the ocean. The

THE BEAUFORT SCALE

The Beaufort Scale uses the effects of the wind on the environment (ocean surface, trees, smoke, and other indicators) to estimate wind speed without the help of a measuring instrument. The original scale, as developed in 1805, was designed to use the wind's effects on a ship's canvas sails as a guide to estimate and standardize wind intensity reports. The specifics of its categories, indicators, and uses have evolved significantly over the years. In the 1930s, ship observers used the state and appearance of the surface of the ocean as a guide to estimate a wind force value from 0 to 12 that could also be described with a qualitative category. Weather Bureau forecasters often used these qualifiers as part of their advisories. The current incarnation of the Beaufort Scale assigns specific wind speed ranges to each category based on wave height, sea surface conditions, coastal conditions, or "land conditions," depending on the specific application. The Beaufort Number or Wind Force 12, described as a "hurricane," does not refer to the type of storm but only to the strength of the estimated wind speed.

Beaufort Scale

Beaufort Number (Wind Force)	Category Description	Estimated Wind Speed (mph)
0	Calm	< 1
1	Light air	1–3
2	Light breeze	4–7
3	Gentle breeze	8–12
4	Moderate breeze	13–18
5	Fresh breeze	19–24
6	Strong breeze	25–31
7	Near gale (High wind, Moderate gale)	32–38
8	Gale (Fresh gale)	39–46
9	Strong gale	47–54
10	Storm (Whole gale)	55–63
11	Violent storm	64–72
12	Hurricane	≥ 73

NOTE: As defined in the World Meteorological Organization International Codes Manual (Volume I.1 Part A).

scale does not allow for distinguishing intensity higher than a minimum hurricane. However, from the low pressure reading, even though it was not precise, we can infer that the actual winds were much stronger than the minimum of near 75 mph, most likely somewhere within the 100–130 mph range, already or soon to become a major hurricane (indeed, HURDAT2 estimates that the storm had sustained winds of 125 mph around this time). The *Alegrete* was well able to withstand the violent weather, but that doesn't mean those aboard would not have experienced a wild ride. Reports from similar ships caught in hurricanes speak of very rough seas, waves clearing the deck, onboard boats smashed, and crew having to hold on with both arms to avoid going overboard.[30] One can only imagine the rough conditions in which the ship found itself and the difficulties in recording the reported weather data.

Modern accounts of the Great New England Hurricane suggest or directly state that on the evening of September 16, the *Alegrete* radioed its report of a hurricane to the Weather Bureau, which quickly got to work monitoring the storm and issuing advisories. The first storm advisory, however, was not issued by the Bureau until the evening of the 17th, which begs the question, Why did forecasters wait a full day before making the public aware of the storm? A close examination of the same abovementioned MWR report reveals that forecasters did indeed send word as soon as they learned about it, but it was not the *Alegrete*'s report that alerted them at first.

> This hurricane was first definitely located from radio reports on the evening of September 17, when it was centered approximately 500 miles northeast of the Leeward islands, but mail reports now at hand show that it was centered at about 21°N, 53°W late on the 16th.

The report then goes on to describe that the Hurricane was not charted before the evening of the 16th and describes the *Alegrete*'s observations. It also directly states that these earliest observations locating a hurricane now "at hand" were obtained via mail, and hence not radioed live as assumed in recent accounts of the storm. The ship that actually provided the first radio report during the evening of the 17th is not identified, but it can be indirectly inferred later in the article: "early on the morning of September 17, the Netherlands' S.S. *Socrates*, [which] encountered the storm while near 21°N., 59°W." The same ship is listed in the "Ocean Gales and Storms" of September 1938 as the "Dutch M.S. Socrates" that experienced a wind force 11 gale during the 17th, with lowest pressure of 29.29 at 10 P.M. and near the

TABLE 3.2. Ships in the Vicinity of the Hurricane

Vessel	Country	Minimum Pressure[a]	Position[b]	Date/ Time[b,c]	Maximum Wind Force[d]
M.S. *Socrates*	Netherlands	29.29	20°38'N, 59°17W	17/10 PM	11
S.S. *Robin Goodfellow*	Trinidad	29.66	21°15'N, 66°20W	18/4 PM	8
S.S. *Pan American*	U.S.	29.69	28°27'N, 69°20W	18/10 PM	8
M.S. *Gulfhawk*	U.S.	29.00	25°31'N, 69°54W	19/10 AM	12
S.S *Jean Lafitte*	U.S.	29.31	27°47'N, 72°53W	20/4 AM	12
S.S. *Antigua*	U.S.	28.24	27°06'N, 73°54W	20/4 AM	9
S.S. *Atlantida*	Honduras	28.14	27°05'N, 74°35W	20/7 AM	12
M.S. *Phobos*	Netherlands	28.81	27°39'N, 73°57W	20/10 AM	12
S.S. *Agwidale*	U.S.	29.64	29°30'N, 72°35W	20/2 PM	10[e]
S.S *India Arrow*	U.S.	28.04	30°00'N, 75°40W	20/6 PM	12
S.S. *Knoxville City*	U.S.	29.69	30°30'N, 72°30W	20/11 PM	11
S.S. *San Benito*	Panama	28.60	36°14'N, 74°36W	21/10 AM	11
U.S.N. *Enterprise*	U.S.	29.39	36°52'N, 75°47W	21/10 AM	10[e]
S.S. *Gulfprince*	U.S.	28.82	38°37'N, 73°54W	21/12 PM	9
S.S. *Platano*	Panama	28.70	38°20'N, 74°56W	21/12 PM	9
S.S. *Birmingham City*	U.S.	28.10	38°55'N, 72°00W	21/1 PM	12
S.S. *R. G. Stewart*	U.S.	28.64	39°18'N, 73°48W	21/1 PM	12
S.S. *Metapan*	U.S.	29.44	40°36'N, 69°26W	21/3 PM	10
S.S. *Washington*	U.S.	28.69	40°28'N, 73°50W	21/4 PM	11[e]

SOURCE: Data obtained from the "Ocean Gales and Storms" table within "Weather in the Atlantic and Pacific Oceans" report in the September 1938 MWR.

NOTE: These are not the only ships that encountered the Hurricane, but those of which the Weather Bureau had knowledge during or shortly after the storm.

[a] Pressure is in inches of mercury, uncorrected for altitude and as reported by each individual ship.

[b] The position and date/time correspond to the minimum pressure observed.

[c] All dates are in September 1938.

[d] The wind force category corresponds to the Beaufort Scale, described on pages 45–46.

[e] Maximum wind category occurred at a different time than at the time of minimum pressure.

same latitudes listed above.[31] There are no other ship reports for that day in the vicinity of the tropics. Therefore, it stands to reason that the *Socrates* was that first ship to send out "live" word of the storm to the Weather Bureau. As shown in Table 3.2, several other ships also had close encounters with the storm during the next several days.

Thunderstorm Spirals and the Eye

Today, constant monitoring of tropical cyclones by meteorological satellites is standard practice. If this important tool had been available in 1938, the group of westward-moving thunderstorms over the African coast would have been identified and monitored closely. Likewise, very soon afterward, the storm would have been seen to develop a rotation and, as it moved westward across the Atlantic Ocean, the appearance of an eye would have eventually become obvious.

It is not until hurricanes intensify and mature that they normally develop the most distinguishing features we associate with them: the cloud-free eye and the spiral bands of clouds arranged in a counterclockwise direction around the center (clockwise in the Southern Hemisphere). The banding and even the eye are not always apparent, because they are obscured sometimes by a shield of high cirrus clouds.[32] Such distinctive structure is not obvious to observers experiencing a hurricane from the ground, as hurricanes are normally hundreds of miles across, with some even extending more than 1,000 miles. When looking at the clouds, the same rotation that is very obvious from the perspective of a faraway meteorological satellite is not at all noticeable.

The understanding that hurricanes are rotating systems, however, did not have to wait until the satellite era. Indeed, it is usually attributed to observations made after the passage of one particular hurricane through New England. Commonly named the 1821 Norfolk and Long Island Hurricane, this storm traveled north along and just off the coast from North Carolina to Maine.[33] The damage in Connecticut (which experienced the passage of the eye of the storm) was carefully investigated by a highly intuitive and inventive man of many interests, a self-taught scientist named William C. Redfield. At that time, Redfield was primarily a shopkeeper, but he later went on to become one of the founders and the first president of the American Association for the Advancement of Science and a well-known storm investigator.[34] The story of how Redfield rode through Connecticut surveying the storm damage has reached somewhat mythical status over the years, and the details vary from account to account (although the actual observations made during his journey were unequivocally recorded). The account written by his great-grandson, Dr. Alfred C. Redfield, is perhaps the most accurate. Dr. Redfield happened to be a senior biologist at Woods Hole Oceanographic Institute in Massachusetts in September 1938, and thus wrote a report as part of the Institute's newsletter issue dedicated to the 1938 storm.

His report included his great-grandfather's story, as told by his grandfather, William Redfield's son:

> My grandfather records in his "Recollections" that on the evening of September 3, 1821, while his stepmother was lying on her deathbed, a gale short in duration but terrific in violence, passed over Connecticut. . . . About a month after this, his father, William C. Redfield, visited Stockbridge [Massachusetts] to carry to his wife's parents some of her belongings and to give the sad history of their daughter's last illness. The journey of seventy miles . . . was made by wagon and occupied two days. As he drove along, he observed that at Middletown and Cromwell [in central Connecticut] the wind had been from the southeast and the trees lay prostrate with their heads northwest. On reaching [western Massachusetts] he was surprised to see that they lay in the opposite direction, and on conversing with the residents of that region, he was astonished to learn that the wind, which at 9 P.M. had been from the southeast at Cromwell, had been in Stockbridge [Massachusetts] from the Northeast at precisely the same hour. These facts at first seemed to him irreconcilable. It did not appear to him possible that two winds of such violence should be blowing directly against each other at the distance of only seventy miles. The only explanation of this paradoxical phenomenon was one which he was then led to accept hypothetically, but which he afterwards confirmed by years of observation and the collection of innumerable facts. . . .[35]

The conclusion reached by Redfield had great implications in our understanding of the nature of hurricanes and storms in general: "There appears but one satisfactory explication of these phenomena. This storm was exhibited in the form of a great whirlwind."[36] Redfield did not publish his ideas right away. He continued developing his theory for a few years and might have kept it all to himself if it wasn't for a chance encounter with a Yale professor who urged him to publish his findings and thoughts on the subject.[37] His first article, "On the Prevailing Storms of the Atlantic Coast," was published in 1831 in the *American Journal of Science and Arts.* Redfield's observations suggested a counterclockwise rotation and the passage of the eye of the hurricane. However, in his account and analysis, following the usage of the time, the term "hurricane" refers more to the strength of the wind ("a wind or tempest of the most extraordinary violence") than to the type of storm itself. He also described counterclockwise motion as "the direction of revolution being from right to left, or against the sun, on the north side of the equator."

Redfield was not, however, the first to notice the rotational nature of these storms (although he was unaware of previous observations), but he was certainly the first to perform an exhaustive analysis of the observations (not just for this hurricane, but for many others that followed) and use them to develop a set of general rules about the characteristics and behavior of storms.[38,39] The rules, soon afterward postulated by Redfield's collaborator, Sir William Reid, as the Law of Storms, represented the first theoretical model to describe not just hurricanes but any type of storm (as the distinction between storms of tropical and nontropical origin had yet to be realized) based on observations. In the early to mid-1800s, Redfield's ideas became the catalyst for heated intellectual debate, mostly between himself and his contemporary, the renowned expert James P. Espy. Completely closed to each other's arguments, both men attempted to discredit worthy portions of their opponent's theories. For example, Espy rejected the rotational nature of storms and Redfield rejected Espy's "caloric heat" (what we would today call "latent heat") theory of storms. Dubbed the American Storm Controversy, the bitter debate went beyond the specifics of the nature of storms themselves and was at its core a debate between different philosophies on how to engage in science: empirically (reaching conclusions from analyzing observations) versus theoretically (reaching conclusions by applying theoretical principles, which at the time would have been indistinguishable from applying philosophical principles).[40]

The eye, a signature feature of many fully developed hurricanes, is the area at the center that is calm and devoid of storm clouds; it is also an area of extremely low atmospheric surface pressure, the lowest in the entire storm. Its surrounding eyewall is made of the highest thunderstorm clouds within the storm (sometimes reaching up to 60,000 feet), and it contains the most violent weather, with the strongest surface winds and heaviest rainfall. Inside the eye, the rain abruptly stops and the winds sharply diminish to calm or very light toward the center. Consequently, even though it might feel and sound somewhat eerie because of the very low atmospheric pressure, the weather experienced at a location during the passage of the eye at landfall could actually be described as good. The shift from the violent wind and heavy rain in the eyewall to the very calm, clearing skies can be quite dramatic. The sky, however, is not always completely clear inside the eye, as some low, fair weather clouds might be present. The previously mentioned September 1938 MWR storm summary, for example, reported that when the "calm center was felt at Brentwood, Long Island between 1:50 P.M. and 2:50 P.M." . . . "[t]he wind movement was so slight during this time that 'a cigarette could

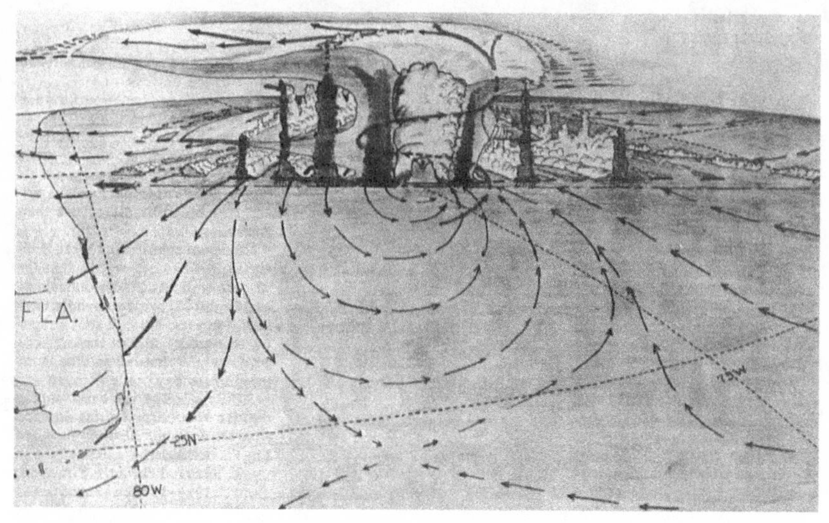

Model of a hurricane's vertical cross-section showing its basic structural features: the eye, eyewall, and bands of thunderstorms. The hand-drawn diagram, appearing in the May 1958 issue of the *Weather Bureau Topics* is based on data collected during field missions as part of the National Hurricane Research Project (NHRP), which was established in 1955 as a response to the devastating 1954 hurricane season. The project flew missions into Hurricanes Greta (1956), Audrey (1957), Daisy (1958), and Helene (1958), for the first time collecting observational data of the structure and energetics of hurricanes. Together with the Miami forecast office and an aircraft facility in West Palm Beach, FL, the NHRP also led to the development of the NHC.

have been lighted in the open without difficulty.'"[41] It is understandable, then, for those experiencing a hurricane and who have no better information at their disposal, to think that a storm is over during the passage of the eye. They might come outside to survey the damage and very soon be surprised when the violent weather returns, as the other side of the eyewall arrives within an hour or so but with winds from the opposite direction.

Indeed, during the early 1900s—a time when the nature of hurricanes was not yet widely understood—after the passage of the eye and as the other side of the eyewall arrived with its accompanying violent weather, people often believed that the storm had shifted course and come back in a sort of boomerang fashion (in the Spanish-speaking Caribbean islands, locals referred to this as "La Virazón," which literally translates to "The Big Turn").[42] A remarkable danger appears when experiencing the eye of a hurricane over the ocean. The sun still comes out, the winds still become light, but the seas are the roughest. With the strongest winds in the eyewall surrounding the center, very large, violent waves come from all directions and the ocean surface becomes very chaotic. These are the conditions that ships would encounter when unknowingly heading straight into the center of a hurricane.

Ship reports would also commonly mention exhausted birds landing on their decks. These birds could not and would not fly through the violent eyewall but would continue flying while trapped inside the eye until they found a place to rest or collapsed in exhaustion. Sometimes they can remain trapped inside the hurricane's eye until it reaches land, providing for some interestingly misplaced birds. The landfall of Hurricane Gloria (1985) on southern New England, for example, was reportedly accompanied by thousands of birds.[43] The same observation was made, albeit not in as dramatic a fashion, after the arrival of the 1938 Hurricane. *The Auk*, a quarterly scientific journal of the American Ornithologists' Union, published a collection of notes reporting various types of tropical and ocean birds spotted in Massachusetts, New Hampshire, and as far away and inland as Vermont. A report from Long Island, besides noting the disappointing lack of exotic birds found in the aftermath of the storm (concluding that they must have been blown inland toward northern New England), remarked how impressive were "the unprecedented flight of [sea gulls] that battled the wind with astonishing success and the flocks of small shorebirds that rather helplessly streaked by" during the height of the storm.[44]

Now a full-fledged hurricane crossing the Atlantic Ocean, the storm that would become the Great New England Hurricane intensified until it finally caught the attention of forecasters. Then, as it drew closer to the Caribbean and Florida, the storm was closely monitored by both Cuban meteorologists and the Weather Bureau Office in Jacksonville. All signs pointed to a monster of a storm imminently hitting Florida.

4
FLORIDA'S SIGH OF RELIEF

Intensification and recurvature: From threatening Florida to speeding northward undetected

Watching and Warning

Official forecasts of track and intensity for Atlantic tropical cyclones as well as advisories with updated storm information for the public are today in the hands of the National Hurricane Center (NHC). The agency is also in charge of hurricane and tropical storm watches and warnings along the coast. All these duties, though more loosely defined than they are today, used to be performed by specifically designated Weather Bureau Offices, whose forecasters monitored the storms and decided when to warn the public (see Chapters 2 and 5 for more on the history of hurricane warning and forecasting).

Although the manner in which hurricane warnings are issued has varied throughout the years, they have always been an announcement of impending hurricane conditions for a location. In the 1930s the decision was rather subjectively made by the "forecaster in charge" of the designated Weather Bureau Office, who would order a warning when he felt that the hurricane was imminent. Warnings often followed, rather than anticipated, the rapid drop in pressure associated with an approaching storm, and it was not unusual for strong winds to already be blowing when the warning announcement was made. Rules specifying the timing of warnings were first introduced a bit later during World War II, when many of the Weather Bureau's resources were tied up in wartime responsibilities. The protocol of the time dictated that hurricane warnings would be issued 24 hours in advance of any potential landfall and only to those areas under direct threat (to avoid giving the enemy useful information about the location of storms at sea).[1,2] The secrecy was, as one would expect, abandoned after the war, but the 24-hour advance warning remained in place for several decades. In 2010, the timing was updated so that a hurricane warning is issued 36 hours in advance of the anticipated onset of tropical storm force winds if hurricane conditions are expected somewhere within the specified area.[3] The change allows adequate time for necessary preparations before the stormy weather begins. A hurricane watch, which is now issued when hurricane conditions are possible within the specified area and 48 hours before tropical storm conditions are expected, did not exist in the 1930s. Its origin dates to 1943, when a "preliminary hurricane alert" to be announced 24 to 36 hours before potential landfall was added to the roster of tropical cyclone advisories.[4] The word *alert* was later dropped in favor of *watch* (which was used for the first time during the 1956 hurricane season), as it was considered less ominous and less likely to cause confusion with military and civil defense alerts.[5,6]

Tropical storm watches and warnings are similarly defined, but for winds 39 to 73 miles per hour (see Chapter 1 for the definitions of various watches and warnings). They started being issued in the 1980s. Previously, when a tropical storm was a threat, gale warnings were issued—in agreement with the "gale" categories in the Beaufort Scale (see Chapter 3) corresponding to tropical-storm strength. In 1938, a gale warning would not have distinguished between a storm caused by a system of tropical origin and one of extratropical origin, the latter of which was much more common in New England and normally of lesser intensity. On September 21, 1938, only gale or whole gale warnings were ordered by forecasters (see Chapter 5 for more information on the warnings issued by the Washington, D.C., Weather Bureau Office in advance of the Hurricane).[7]

As discussed in Chapter 2, the hurricane monitoring, forecasting, and warning duties of the Weather Bureau in the 1930s were collectively known as the Hurricane Warning Service. The system in place in 1938 had only been established three years earlier. Prior to that these duties were carried out only by the Washington, D.C., Office. Since 1935, designated Weather Bureau locations would be in charge of tracking, forecasting, and informing the public while a storm was located within its assigned geographical region. San Juan, Puerto Rico, was responsible for the Caribbean Sea and any islands east of 75°W and south of 20°N; New Orleans for the portion of the Gulf of Mexico west of 85°W; and Jacksonville, Florida, for the remaining portion

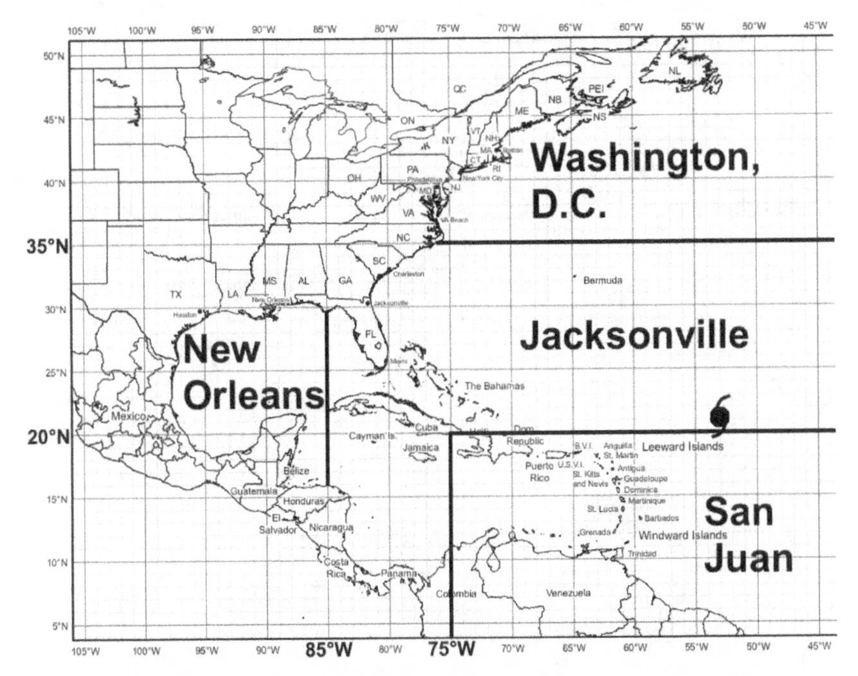

The earliest observed position of the Hurricane was near 21°N, 53°W (marked with a hurricane symbol), which meant that the Jacksonville Weather Bureau Office was in charge of its monitoring, forecasting, and issuing of advisories and warnings. After the storm moved north of 35°N, its responsibility passed to the Washington, D.C., Office. (Base map from the NOAA NHC Atlantic Basin Hurricane Tracking Chart)

of the Atlantic coast, Caribbean Sea, and Gulf of Mexico south of 35°N. The Washington, D.C., Central Office was responsible for storms north of 35°N (once they were past Cape Hatteras, North Carolina).[8]

Florida Awaits

When the Weather Bureau first learned about the storm that would become the Great New England Hurricane, its position north of 20°N, south of 35°N, and east of 85°W meant that it fell under the Jacksonville Office jurisdiction. At this time, it was nearing its peak of intensity and appeared to be heading directly for Florida. Of course, Floridians were no strangers to devastating tropical systems. The HURDAT/2 contains more than 100 hurricanes making landfall somewhere in Florida since 1851, with more than 30 being of major intensity (category 3 or higher) and almost 20 of those major storms occurring before 1938.[9] Very recently, the Florida Keys had been flattened by the most intense hurricane to have ever made landfall on U.S. soil to date.

THE LABOR DAY HURRICANE OF 1935

The Labor Day Hurricane of September 2, 1935, was the first of only three storms to have made landfall as category 5 hurricanes in the United States. (The other two were Camille [1969] and Andrew [1992].) The storm roared through the Florida Keys with estimated winds between 150 and 200 mph (185 mph estimated landfall speed in the HURDAT/2 database) and a storm surge of 18–20 feet, killing hundreds. The storm hit after the reorganization of the Hurricane Warning Service and the appointment of Grady Norton as chief forecaster in the Jacksonville, Florida, Weather Bureau Office. His issuing of hurricane warnings more than 12 hours in advance of the storm was considered a success, given the limited tropical cyclone observation capabilities of the time. Public reaction to the deadly disaster prompted Congress to appropriate additional money to improve hurricane observation and report capabilities later the same year.

The 1935 Labor Day Hurricane resulted in more than 400 fatalities, famously including more than 250 World War I veterans.[10] Previously unemployed, they had been hired by the Works Progress Administration (WPA) to build U.S. Highway 1. The WPA—one of Franklin D. Roosevelt's New Deal federal agencies—employed millions to carry out public projects in an effort to ease the high unemployment resulting from the Great Depression.[11] The Florida East Coast Railroad had dispatched an 11-car train to evacuate the veterans, but it was swept off the tracks by a 20-foot storm surge and never made it to its destination.

It is not surprising, then, that when the Weather Bureau advised Florida residents of a threatening hurricane just three years later, they took it very seriously. People boarded up their homes and businesses and stocked up on supplies in anticipation of the storm. Furthermore, power and telephone workers and Red Cross officials headed south toward the expected storm, armed with tools and lessons learned from previous storms to assist in the expected aftermath and recovery.[12] The Weather Bureau Office in Jacksonville and its forecasters were also ready, paying very close attention and even advising the local population of the possibility of upcoming hurricane warnings.

Immediately after the Hurricane, the Secretary of Agriculture requested an explanation of the events that transpired, and two weeks later the Weather

Bureau complied by sending a report that included, among other things, a list of all the "Advisories Issued in Connection with the Tropical Hurricane of September 17–21, 1938."[13] Jacksonville circulated a total of 15 advisories from September 17 to September 21: eight while the storm was moving toward Florida, three more as it was starting to turn, and four more after it was heading in a more northward direction, hinting at a northeastward turn. Careful examination of these advisories and those later issued by Washington, D.C., give us valuable insights into how the hurricane threat was handled at the time.

The text of the first advisory, which as previously mentioned was transmitted as soon as the Weather Bureau had word of the presence of a hurricane, partially reads as follows:

> Advisory nine thirty P.M. [September 17] A well developed tropical disturbance probably of full hurricane intensity has appeared some five hundred miles northeast of the Leeward Islands . . . apparently moving westward about fifteen to eighteen miles per hour. Caution advised all vessels in path of this dangerous storm.

The subsequent three advisories the next day continued cautioning all vessels, specifically instructing "all small craft Cape Hatteras to Florida Straits" to navigate carefully until the storm danger passed. The first mention of potential danger for Florida came later that evening:

> Advisory nine thirty P.M. [September 18] Hurricane centered . . . about nine hundred miles east-southeast of Miami moving west-northwestward about twenty miles per hour attended by strong shifting gales and squalls over large areas and by hurricane winds near center. Caution advised all vessels in path and all small craft Hatteras to Florida Straits should remain in port until storm danger passes. Storm will begin to affect extreme eastern Bahamas by midday Monday and central Bahamas by Monday night. Interests on east Florida coast should keep closely in touch with further advices. If present direction and rate of movement is maintained winds will begin to increase on east Florida coast early Tuesday.

A strong gale has a very specific meaning in the Beaufort Scale. It corresponds to a wind force 9, which translates into 47 to 54 mph or moderate tropical storm intensity winds on the outskirts of the storm accompanying the hurricane intensity winds in the center.

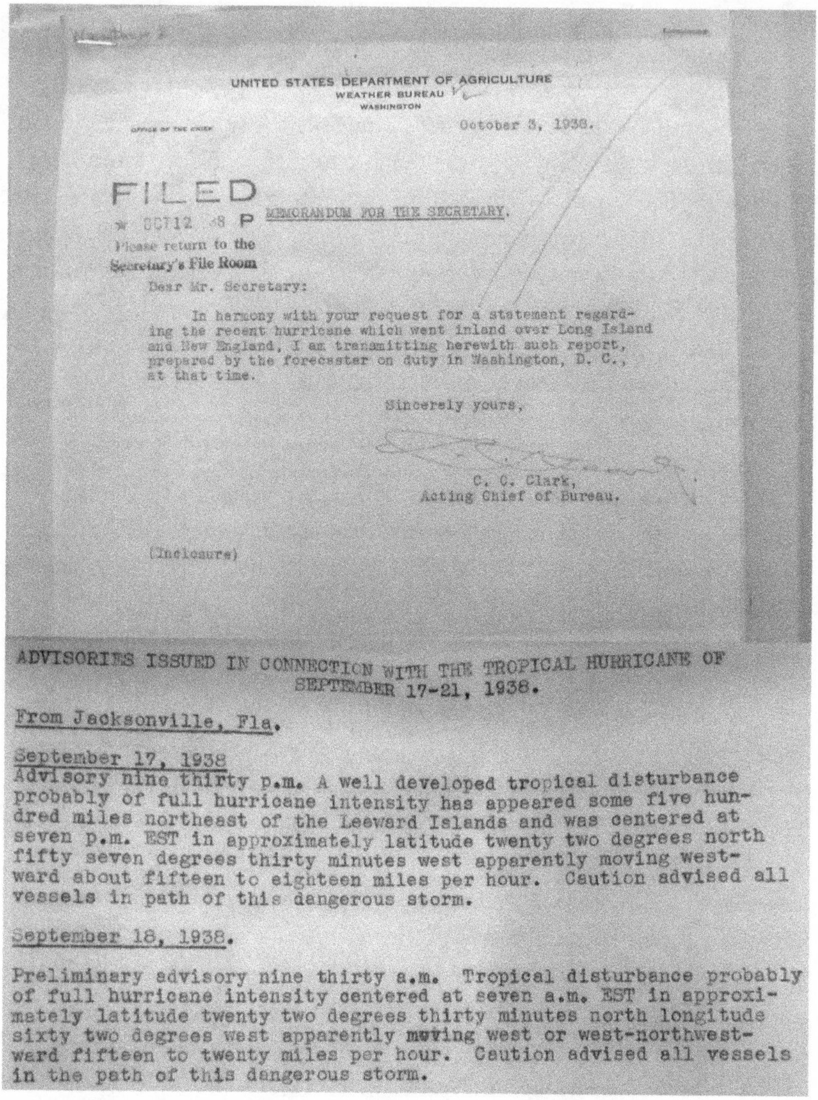

UNITED STATES DEPARTMENT OF AGRICULTURE
WEATHER BUREAU
WASHINGTON

OFFICE OF THE CHIEF

October 3, 1938.

FILED
OCT 12 38 P
Please return to the
Secretary's File Room

MEMORANDUM FOR THE SECRETARY.

Dear Mr. Secretary:

In harmony with your request for a statement regarding the recent hurricane which went inland over Long Island and New England, I am transmitting herewith such report, prepared by the forecaster on duty in Washington, D. C., at that time.

Sincerely yours,

C. C. Clark,
Acting Chief of Bureau.

(Inclosure)

ADVISORIES ISSUED IN CONNECTION WITH THE TROPICAL HURRICANE OF SEPTEMBER 17-21, 1938.

From Jacksonville, Fla.

September 17, 1938

Advisory nine thirty p.m. A well developed tropical disturbance probably of full hurricane intensity has appeared some five hundred miles northeast of the Leeward Islands and was centered at seven p.m. EST in approximately latitude twenty two degrees north fifty seven degrees thirty minutes west apparently moving westward about fifteen to eighteen miles per hour. Caution advised all vessels in path of this dangerous storm.

September 18, 1938.

Preliminary advisory nine thirty a.m. Tropical disturbance probably of full hurricane intensity centered at seven a.m. EST in approximately latitude twenty two degrees thirty minutes north longitude sixty two degrees west apparently moving west or west-northwestward fifteen to twenty miles per hour. Caution advised all vessels in the path of this dangerous storm.

A letter (above) dated October 3, 1938, from Acting Weather Bureau Chief Charles C. Clark to the Secretary of Agriculture, Henry A. Wallace, in response to a "request for a statement regarding the recent hurricane that went inland over Long Island and New England" contained a full report of the Bureau's actions during the storm and a listing of all the advisories issued in connection with the storm (below). The text of the entire report is included here as an appendix. The original document is in the Department of Agriculture textual documents at the National Archives and Records Administration at College Park, MD.

The September 19, 3 A.M., advisory had more urgent words, stating that "all interests in [the] path of this severe storm should exercise extreme caution" and that "Florida is in the danger zone of this storm and all persons are urged to stand by for later announcements today." The 9:30 A.M. advisory, also on the 19th, for the first time had specific information about warnings. It stated that northeast storm warnings were ordered for Jacksonville to Key West, with "northeast" referring to the wind direction expected in the area during the storm. This type of warning is not specific to hurricanes and was commonly used for other less severe coastal storms. This was especially true for areas farther north, where a "northeast storm warning" might be issued several times a year during nor'easters (offshore extratropical cyclones for which the New England region experiences northeast winds during the most intense weather). The 9:30 A.M. advisory also urged all interests on the Florida east coast to "stand by for possible hurricane warnings during the day."

The 3 P.M. advisory contained even more urgent language in its message to Florida residents:

> Advisory three P.M. [September 19] Hurricane centered . . . about five hundred and thirty miles east-southeast of Miami apparently still moving west-northwestward nearly twenty miles per hour attended by gales and squalls over a large area and hurricane winds near center. If present direction and rate of movement is maintained center of storm will . . . reach southeast Florida coast in twenty four to thirty six hours with winds beginning to increase on coast late tonight. All interests in southern Florida should immediately make all possible preliminary preparations to withstand this severe storm and then stand by for later advices. Hurricane warnings probably will be issued tonight. Northeast storm warnings remain displayed Jacksonville to Key West.

This is roughly the time when a hurricane warning would have been issued under current guidelines, a stark contrast to the practices in the 1930s when, as previously described, warnings were not announced until the storm was imminent.

As part of the reorganization of the Hurricane Warning Service in 1935, it was ruled that "the name of the forecaster will be omitted from all forecasts, warnings, and advisories relating to storms and hurricanes; only the signature 'Weather Bureau' will be used on such message for general distribution."[14] One can only speculate about the motivation behind this policy, but as a result, the names of the forecasters responsible for each of the advisories from Jacksonville and later from Washington, D.C., were not officially recorded. Nevertheless, history would not let them stay anonymous. The same two men that dealt with the 1935 Labor Day Hurricane in their duties as part of the newly established Jacksonville Weather Bureau Office were also in charge of monitoring the Great Hurricane of 1938.[15] Experienced forecaster Grady Norton and his younger assistant, Gordon Dunn, spent five sleepless days and nights carefully watching the storm and doing an excellent job of keeping the residents in the threatened area informed. Both men would continue lifelong careers in the field. Norton, who was considered by his contemporaries as having the ability to "smell" the storms, became the most trusted hurricane forecaster of his time, recognizable as the reassuring voice providing timely information about threatening storms on the radio and even through public address systems in grocery stores, restaurants, and other establishments in the Florida area.[16] He continued forecasting hurricanes until October 9, 1954 (during the same infamous hurricane season that led to the creation of the NHC), when he died suddenly after completing, against his doctor's advice, a 12-hour shift during Hurricane Hazel.[17] His exemplary hurricane forecasting career was recognized and celebrated in articles appearing in the *Weather Bureau Topics* both right before and right after his death.[18,19] Gordon Dunn was, as it turns out, the first to identify the small pressure disturbances traveling westward over the Atlantic and later named easterly waves (see Chapter 3 for more on easterly waves).[20] He was also the first to notice that the formation of the majority of hurricanes over the region was associated with such disturbances. He succeeded Norton as chief hurricane forecaster in Miami (where hurricane warning operations had been moved from Jacksonville in 1943), becoming the first director of the NHC—though as the Miami Office director, Norton was posthumously recognized by many as the first director of the agency.[21] Dunn went on to receive many awards for his efforts at the NHC, including the Department of Commerce Gold Medal in 1959.[22]

Hurricane Maintenance and Intensification

Just as the storm was starting to be watched from Florida and the Caribbean on September 17, it was strengthening into a major hurricane. The storm never weakened below the strength of a category 3 storm until after landfall on the afternoon of the 21st.

For a hurricane to intensify and maintain its strength, it needs a nurturing environment that provides enough energy and at the same time does not disrupt the storm's self-sustaining mechanisms. These needs have tra-

TABLE 4.1. Classic List of Environmental Conditions Favorable for Hurricane Formation, Maintenance, and Intensification

Warm ocean water	At least 80°F (26.5°C)
	To a depth of about 50 m (150 ft)
Moist mid-troposphere	High relative humidity
Unstable Atmosphere	An environment cooling with height fast enough such that rising air will still be warmer than its surroundings and continue rising—and have the ability to build large thunderstorm clouds
Preexisting disturbance	With enough background rotation
	With enough low-level inflow of air
Weak vertical shear	Less than 20 kt (25 mph) difference between upper and lower winds
Enough distance from equator	At least 5° away (500 km, 300 mi)

ditionally been expressed as a short list of favorable environmental conditions, which are necessary but not sufficient.[23] In other words, a suitable environment does not guarantee a hurricane. On the flip side, for a storm to reach category 4 or 5 intensity, conditions in their most favorable form are certainly needed.

The undisputable source of power for a hurricane is the energy that it obtains from the ocean, which directly and indirectly provides huge amounts of heat to the air. This heat is, in turn, used to power the storm. As air spirals toward the center, it ventilates the surface of the ocean and collects the heat and water vapor that evaporates from the surface. The indirect source of heat (known as "latent heat") is provided by the water vapor when it condenses to form cloud droplets (as the change from the more energetic gaseous state to the less energetic liquid state releases the excess energy as heat). This latent heat is commonly and mistakenly thought to be the driving force behind the hurricane's energy, but it turns out that it is not as important as the total heat fluxes from the surface of the ocean.[24] The additional heat (regardless of its direct or indirect origin) increases the buoyancy of the air, which in turn causes the air to further accelerate toward the center, collecting even more heat from the ocean, further increasing the buoyancy and so on. The self-intensifying mechanism set in motion can continue as long as the center of the hurricane remains over the warm water. One can think of a mature hurricane as an engine converting the heat energy provided by the ocean to the mechanical energy of the hurricane winds. This analogy is not only qualitative, but corresponding mathematical calculations can be performed to determine the maximum possible intensity achievable for a hurricane passing over an ocean surface of a given temperature.[25]

By analyzing data from around the world, tropical meteorologists have determined that there appears to be a threshold sea surface temperature (SST) of 80°F (26.5°C) below which tropical cyclones tend not to form.[26] This threshold was first proposed in 1948 by Finnish meteorologist Erik H. Palmén and substantiated in 1968 by William M. Gray, now well known for his group's much-anticipated and discussed seasonal hurricane predictions.[27,28] Gray's landmark comprehensive analysis of tropical cyclones around the world led to the formulation of the well-known list of necessary environmental conditions for hurricane genesis, intensification, and maintenance shown in Table 4.1.[29] The handful of storms observed to have formed over ocean waters below 80°F are normally aided by a nontropical energy source such as an old frontal boundary with marked temperature contrasts.[30] SST does not vary much from day to day, meaning that its effects are more of a seasonal nature. In fact, the time of year when tropical cyclones tend to form around the world correlates well to the seasons with warm enough ocean surface temperature.[31] This is why Cape Verde hurricanes (see Chapter 3), such as the 1938 storm, tend to form only from the middle of July to late September, when SST is normally warm enough throughout the tropical corridor from the African western coast to the Caribbean.[32] As one would expect, the warm SST requirement also means that hurricanes weaken and eventually dissipate when moving onto land or when reaching cooler oceanic regions.

Another important environmental condition is the depth of the warm water. The top layer of the ocean is normally a well-mixed, constant temperature layer below which water temperature drops very rapidly with depth. When a hurricane's winds stir the surface, they cause water from below to be brought upward and mix with the surface water. For the stirring not to cause cooling of the SST to unfavorably low temperatures, the warm layer has to be sufficiently deep. Eighty degree Fahrenheit water at a minimum depth of about 150 feet (or 50 meters) is a good rule of thumb for an environment where a hurricane's development will not be impeded by the intrusion of colder water from greater depths.[33] SST is thus also affected by the hurricane itself. The stirring of water by a strong hurricane always causes a certain degree of cooling, especially near the strongest winds toward the center of the storm. It is, therefore, not unusual for a relatively narrow trail of cooler water to be caused by intense hurricanes and be seen in satellite SST imagery. Given its maximum category 5 strength, it is very possible that the 1938 Hurricane would have left such a cool trail behind during part of its track over the ocean.

A hurricane rarely, if ever, reaches its potential maximum intensity. Environmental conditions other than SST constraints, such as wind shear (the change of wind speed and direction with height), also put a limit on how much a storm can intensify. For the heat obtained from the ocean surface to be used most efficiently by a hurricane, it must be kept in the core of the storm. This is not possible if the mechanism is disrupted by winds blowing at different speeds or directions above the surface. Therefore, weak wind shear is needed for intensification. This requirement was first recognized in 1944 by Herbert Riehl (the same developer of the easterly wave model described in Chapter 3).[34] Wind shear relevant to tropical cyclones is usually measured by comparing the environmental wind speed in which the hurricane is moving in the upper troposphere with that close to the surface.[35,36] Unlike SST there is not a recognized maximum quantitative threshold, but a difference of less than 20 knots (roughly 25 mph) between the upper and lower wind speeds is generally considered favorable enough to maintain a storm. Because low-level winds in the deep tropics tend to be light, a requirement of weak wind shear means that upper winds also need to be light. Consequently, this condition is sometimes expressed as "light upper-level winds" instead of "weak wind shear."

The weaker the shear, the better—that is, unless the winds steering the storm are so light that the storm is barely moving, which could cause it to interact with the colder surface waters produced by its own stirring. Additionally, to make matters more complicated, wind shear does not affect all storms in the same way.[37] Well-established larger and stronger hurricanes are not influenced as much or as quickly by shear—the effects are stronger on weak or developing systems. Wind shear must have certainly been weak for the 1938 Hurricane to form and continue to intensify during its westward journey. Then, as the storm started moving toward higher latitudes, wind shear would have increased (since upper winds become stronger), but by then the storm was already a well-established major hurricane, and therefore harder to disrupt.

Tropical ocean surface air is laden with moisture, because generally speaking, evaporation from the warm ocean surface provides the atmosphere with an almost infinite source of water vapor. At middle levels in the troposphere, on the other hand, tropical air can sometimes be quite dry, which of course would disrupt the formation or maintenance of a storm. The

Sahara Desert in northern Africa serves as a huge source of dry air known as the Saharan Air Layer (SAL).[38] This mass of dusty air can sometimes reach well westward onto the tropical Atlantic, sometimes as far as the Caribbean Sea and Florida. A disturbance cannot develop into a tropical cyclone inside this dry air and any previously formed hurricane is severely affected if infiltrated by it. (Interestingly, the SAL can also be associated with enhanced areas of background rotation to its south, providing potentially favorable pockets for tropical cyclone formation.)[39] There is no indication that the New England Hurricane was affected by any intrusion of dry air during its westward journey.

The last necessary condition is not environmental per se, but a characteristic of the planet itself and directly correlated with latitude. Briefly introduced in Chapter 3, the Coriolis effect or Coriolis force is a result of the planet's rotation and causes the air to deflect toward the right of its motion in the Northern Hemisphere. To understand how this effect is related to the rotation of a mature hurricane, its interaction with the other forces that also affect atmospheric motions must be examined. The fundamental force that sets the air in motion is known as the pressure gradient force. Areas of high pressure weigh more per unit area at the surface, which means that there is a larger amount of air above them or that the air is denser when compared to an equivalent column overlying an area of lower pressure. The atmosphere tries to adjust for any existing horizontal density differences and essentially pushes the air away from higher and toward lower pressure. If Earth did not rotate, this is exactly how air would always move: directly from high to low pressure. However, it would be very difficult to maintain (or even form) an intense low pressure system, much less a hurricane, since air would forcefully fill the center of the storm, increasing the weight and therefore the pressure. Of course, Earth does rotate, and even though some air does move into the center of surface low pressure systems, the amount is limited by the flow that results when all the forces come into play. The force balance achieved inside the storm includes the Coriolis and pressure gradient forces, as already described, as well as the outward push on the air when it moves in circular fashion, or the centrifugal force.[40,41] The result (obtained from adding all force vectors) is a counterclockwise flow around low pressure systems in the Northern Hemisphere (clockwise in the Southern Hemisphere). Toward the surface, friction causes the flow to slightly cross toward the center of a general low pressure system. For a hurricane, the resulting horizontal flow is an inward spiraling of air that becomes much tighter toward the center. The dominant forces in this balance change dramatically throughout an intense storm. Toward the center of a hurricane, more specifically within the eyewall, where the pressure gradient is the largest and the wind speed is the strongest, the centrifugal force is so strong that it almost fully balances the pressure gradient force. Toward the outskirts of the storm, where the turning of the winds is not as violent, the centrifugal force is much smaller and the Coriolis force is the one dominating the balancing of the pressure gradient. In between these two extremes, throughout most of the storm, all three share in the balancing act.[42,43]

The Coriolis effect is three-dimensional. Whether air is moving horizontally or vertically in relationship to Earth's surface, it will be affected (although there is no horizontal Coriolis force at the equator or vertical Coriolis force at the poles). However, when considering the flow from the outskirts (higher pressure) to the center (lower pressure) of a hurricane, only the horizontal effects need to be taken into account (and the same is true for most meteorological applications). Because the horizontal component of the Coriolis force is zero at the equator, tropical cyclones never form there.[44,45] Even outside the equator, the deflection is very small in the deep tropics where hurricanes form. At very low latitudes too close to the equator, no matter how strong a tropical disturbance is, it will never develop a significant sustainable rotation. Indeed, tropical cyclones do not normally form inside five degrees north or south of the equator, although there have been a handful of exceptions.[46] The 1938 Hurricane and its initiating disturbance were well outside this unfavorable region, never closer to the equator than 12.5 degrees of latitude.

Hurricane experts do not yet fully understand why a disturbance in favorable conditions might never develop a rotation and intensify. It may be that the specific energetic profile of the disturbance has a lot to do with it. Easterly waves that later produce tropical cyclones might possess favorable characteristics even before they leave the African continent.[47] In the case of the 1938 Hurricane, a tropical cyclone did obviously form, implying a favorable initiating disturbance, the easterly wave observed at Bilma on September 4. Additionally, the time of year favored warm enough SST in the tropical Atlantic to provide the needed energy, and the track was far enough away from the equator to allow the initial development and maintenance of a rotating storm. Furthermore, since the storm was intact and intensifying for several days, the environment in which it moved most likely also contained plenty of moisture in the middle atmospheric levels as well as weak wind shear. The environment likely continued to be highly favorable throughout the westward path across the Atlantic Ocean.

The Northward Turn

One might argue that Jacksonville did "too" good of a job with their advisories, since most ships stayed in port afterward, and the number of weather observations and general sea surface conditions being relayed from the ocean dropped significantly as a result. This caused subsequent estimates of the position and intensity of the storm to be less accurate. Indeed, the previously mentioned report from the Chief of the Weather Bureau to the Secretary of Agriculture argues that the storm's predicted movement "was as accurate as could reasonably be expected with no observational data over a wide area to the northeast, east and south of the storm center." Many hurricane forecasters faced this problem before meteorological satellites and aircraft reconnaissance became commonplace: hurricane monitoring was greatly dependent on the reports relayed by ships, which for obvious reasons would do everything possible to get away from the path of a storm once its presence was known.

Regardless of how accurate the knowledge of the storm's whereabouts, it did eventually become obvious that the Hurricane was now going to miss Florida. The first sign of this possibility, though still cautiously stated, appeared in a special early evening advisory:

> Bulletin six thirty P.M. [September 19] Barometer readings in Bahamas since one P.M. have shown only gradual fall in pressure with lowest pressure reported twenty nine sixty inches on Cat Island. At five P.M. EST slow fall in pressure this region indicates storm may be turning towards the northwest. However, interests on southeast Florida coast urged not relax vigilance until recurving tendency is definitely established.

Because barometric pressure was not decreasing rapidly even as a powerful hurricane with very low central pressure approached forecasters were forced to consider a shift from their previous assumptions. The storm was either weakening very rapidly (less likely at this time) or not moving in that direction at the same rate as it was before. This suggested the possibility of a change in direction, in this case from westward to northwestward—the beginning of the hurricane's recurvature.

Semi-permanent, large-scale wind and pressure distributions around Earth, on average, follow predictable patterns consistent with what is known as the general circulation of the atmosphere. The westerly surface winds in the middle latitudes (that normally bring weather from west to east) and the easterly trade winds in the tropics (that normally bring hurricanes and other tropical weather from east to west) are part of this general circulation. High pressure systems over the subtropical oceans, centered roughly around 30 degrees north and south of the equator, are also features of this large-scale atmospheric flow.[48] Air tends to flow in a clockwise direction around centers of high pressure.[49] The steering currents associated with the Bermuda High in the central north Atlantic Ocean usually cause hurricanes to start turning northwestward, then northward, and if not interrupted by the North American continent, then northeastward after their initial westward movement across the Atlantic.[50] This turning is known as the recurvature of a hurricane. A common track of many Atlantic hurricanes is, therefore, to recurve all the way onto the northern regions of the Atlantic Ocean, missing the U.S. East Coast. In September 1938, once it was obvious that recurvature had begun, forecasters assumed that the Hurricane would head toward the North Atlantic and weaken in the process—as it encountered strong wind shear and ocean waters too cold to sustain it. Of course, this Hurricane would not continue its recurvature. Instead, it would keep moving northward toward Long Island and New England (see Chapters 5 and 6 for more about the meteorological reasons for the storm's northward track).

The 9 P.M. advisory on the same day contains language specifying in no uncertain terms that recurvature was indeed happening, though it still advised caution:

> Advisory nine P.M. [September 19] Hurricane centered at seven P.M. EST" . . . about four hundred twenty miles due east of Miami attended by gales and squalls over large area and by hurricane winds near center. Storm has turned northwestward and will provably recurve north-northwestward or northward next twenty four hours. Storm threat to Florida east coast has greatly diminished although interests this area should follow advices carefully next twelve hours. Vessels in path of this severe storm should exercise extreme caution.

In the end, actual hurricane warnings were never needed for the Florida coast. The 9:30 A.M. advisory on September 20 was free of cautionary words and described the expected recurvature and passage east of Cape Hatteras sometime that evening. Preparations were disassembled and life went back to normal. Newspapers declared the excellence of the Weather Bureau for keeping Floridians safe and everyone in the originally expected path of the storm finally exhaled a huge sigh of relief, as indicated by a small and somewhat poetic editorial piece titled "HURRICANE" on page 24 of the *New York Times*: a hurricane that rushed "to strike terror into the hearts of Floridians and remind, even far-away New York, that nature is not to be trifled with when she is in one of her angrier moods."[51] It then goes on to mention that the cyclone "happily spared our Southern coast." Finally, it states that "if New York and the rest of the world have been so well informed about the cyclone it is because of an admirably organized meteorological service." Sadly, this short piece might have been read by some of the same New York residents in Long Island who would lose their homes, loved ones, or even their own lives later that same day.

Nonetheless, for now there was no sign of the impending tragedy. After the storm no longer threatened Florida, it tracked northward and even turned slightly northeastward, staying off the coast and appearing to weaken in the process, all as expected. The last advisory from Jacksonville came very early in the morning of the day of landfall. It suggests that forecasters did not expect the storm to even be north of Cape Hatteras until the afternoon:

> Advisory three A.M. [September 21] Hurricane central . . . about two hundred and twenty five miles south of Cape Hatteras moving rapidly north or possibly slightly east of north. Indications are that center will pass near but slightly off the Carolina Capes within the next twelve hours attended by dangerous gales and high tides on the coast and by hurricane winds short distance off shore. Storm warnings are displayed north of Wilmington North Carolina to Atlantic City, New Jersey. Caution advised ships in the path of this severe storm.

There was no further mention of recurvature or an eventual northeastward track after passing Cape Hatteras by Jacksonville. This is reasonable, since that was outside their responsibility (and inside the D.C. Office's). The 3 P.M. advisory on the previous day, September 20, had mentioned warnings between Cape Hatteras and Wilmington and the 9:30 P.M. advisory, also on the 20th, was the first to mention the warnings between Wilmington and Atlantic City, but saying that they "remain displayed." These warnings along the coast, which had been ordered by the Washington, D.C., Weather Bureau Office, were generic storm warnings and did not acknowledge the presence of a hurricane. The D.C. Central Office took full responsibility for the storm once it became clear that it was past Cape Hatteras, issuing their first advisory at 9 A.M. Fair or not, the decisions made by the D.C. Office later that day would infamously pass into history as a failure of the Weather Bureau's Hurricane Warning Service.

5 "IT DOESN'T HAPPEN IN NEW ENGLAND"

The actions and inaction of the Washington, D.C., Weather Bureau Office and the men that forecasted during the Hurricane

The U.S. Weather Bureau's Central Office was housed in the "Old Ferguson Building," near the intersection of 24th and Northwest M Streets in Washington, D.C. Originally built as a diplomatic residence, the castle-looking mansion served as Weather Bureau headquarters for several decades (from the late 1880s to the early 1940s).[1] This remarkable building was the setting in which significant events unfolded on September 21, 1938, when early in the morning the responsibility for forecasting the Hurricane had fallen onto the lead D.C. forecaster and his staff.

The Weather Bureau in Transition

In 1938, the Weather Bureau was in the midst of a transition from the "old ways" to the "new ways." Since its creation in 1870 as a military division, the weather services agency of the United States had undergone tremendous growth. The number of weather observing stations, the number of men involved in observing and predicting the weather, and the number and scope of obligations and expectations had all increased exponentially. The numerous duties are well summarized as part of the Act of Congress of October 1, 1890—when the agency became a civilian endeavor—under the responsibilities of the Chief of the Weather Bureau, who:

> shall have charge of the forecasting of weather; the issue of storm-warnings, the display of weather and flood signals for the benefit of agriculture, commerce, and navigation; the gauging and reporting of rivers; the maintenance and operation of sea-coast telegraph lines and the collection and transmission of marine intelligence for the benefit of commerce and navigation; the reporting of temperature and rainfall conditions for the cotton interests; the display of frost and cold-wave signals; the distribution of meteorological information in the interests of agriculture and the commerce, and the taking of such meteorological observations as may be necessary to establish and record the climatic conditions of the United States, or as are essential for the proper execution of the foregoing duties.

In just 20 years, the responsibilities of the agency had grown from the seemingly simpler (albeit groundbreaking at the time) simultaneous observations and dissemination of weather conditions to an overwhelming slew of applied tasks.

Once the usefulness of organized weather reporting and forecasting was recognized, all agencies, businesses, trades, political units, and many others jumped on the bandwagon of demanding more customized forecasts and applications. This increase in demand continued for the following four decades, well into the 1930s. One of the latest additions, support for airplane navigation, had grown so large since the 1920s that it was taking over an incredibly large percentage of the agency's resources, costing about half as much as all other services combined (including the immensely important general forecasting and warning activities).[2] Congress appropriations had not grown at a corresponding pace, even having diminished during the heart of the Depression. As a result, then-Chief of the Weather Bureau Willis R. Gregg found himself in a tight situation where the resources at his disposal were being spread thinner and thinner by increasing demands and a decreasing budget.

Knowledge of weather conditions was immensely important to aviation. Flying with the help of onboard instrumentation was still in its very early days and pilots depended on visual cues and good weather for successful and safe flying. From the Bureau's perspective, its involvement in aviation support activities turned out to be both a curse and a blessing. It demanded huge resources: additional weather observation stations at airports, additional personnel to man those stations, and more importantly, increased knowledge of the weather in the atmosphere above the ground, then commonly known as "the free air" (now commonly referred to as the "upper air").

The building on 2416 Northwest M Street does not stand anymore. It housed the Weather Bureau's administrative offices, various research and service offices, and the most important forecasting office for the nation (from *Weather Bureau Topics, Fiftieth Anniversary Issue* [1941]).

In its evaluation of the Weather Bureau, the Science Advisory Board (SAB; see Chapter 1) had emphasized the importance of upper-air observations in forecasting. The initial establishment of such observations, however, was driven by the need to know weather conditions at flying levels. As discussed in Chapter 2, measurements were at first obtained by myriad methods: some by pilot balloon, some by tethered kites, and just a few by fully equipped weather balloons. The majority of the most useful observations at the time, however, were made by especially instrumented airplanes. These early measurements of the free air often were not coordinated but rather obtained as needed or as made possible by the weather conditions themselves. Furthermore, the main objective in using these observations was to brief local and incoming pilots on the current conditions through which they would be flying rather than to use for forecasts to the general public. The SAB's urging for upper-air observations in order to see a summarized picture of the atmosphere above—to be able to draw upper-air maps—made the need for coordination even more essential. So it was that the initial observational demand and drain of forecasting resources in the aid of aviation eventually translated into the development of more sophisticated and accurate forecasting techniques.

Besides the general forecasting advances this new information provided for, the knowledge of air motions above the surface is essential in forecasting the track of a hurricane. To predict whether a recurving hurricane will hit Florida, the Carolinas, or the Northeast, the actual steering currents in which the storm is moving, or the "general drift of the atmosphere," as it was then described, must be known. At the time, it was only possible to know what the currents were at the moment—which was as far as the technology would allow, if only marginally, during the early 20th century—never mind the forecasting of what those currents would do within the next few days.

By the early 1930s, forecasting methods had not changed much since professor Cleveland Abbe started preparing his "probabilities" during the Signal Service years.[3] A careful application of what we would now call "persistence and climatology forecasting" was the most useful tool at hand. Thanks to the telegraph, it was possible to know what the weather was doing upstream (generally to the west) and get a rough estimation of how fast and in which direction those weather conditions were moving across the country—this would be the persistence method. Once enough years of data were gathered, the climatological efforts of the Bureau (in other words, the determination of normal or average temperature and precipitation as well as the "extremes," or highest and lowest, observations for a certain day or time of the year) also gave forecasters a good feel for what type of weather was possible in a certain location or region—this would be the climatology method. These techniques might seem somewhat primitive compared to early-21st-century capabilities, but even attempting this kind of forecasting at the time was an impressive feat, especially for such a large country as the United States and without the use of computers. As one would expect, these methods were more successful at certain times than others and weather forecasting was a much less accurate enterprise than it is today; still, it provided valuable help to farmers, businesses, and even individuals simply needing to plan their day's activities.

The report of the SAB committee did not just contain the general recommendation to start implementing the new airmass analysis techniques to day-to-day weather forecasting but also included very specific guidelines on how to go about it: the types and numbers of upper and surface observations to record and the number of maps that should be constructed and how often; the ways to train existing forecasters and new recruits; and even the way to organize the coordination of all the efforts.[4]

The terminology "airmass analysis" is not used anymore, and it might give those who are familiar with the definition of an air mass (a very large body of air with similar temperature and moisture characteristics) the impression that the location and motion of masses of air are determined by surface observations alone. In a report for *Science*, Chief Gregg, who directed the Bureau from the time of the report's recommendations to the week before the 1938 Hurricane (see Chapter 1), described airmass analysis along these simpler lines:

> . . . air mass analysis consists of a detailed study of masses of air of decidedly different structure as to temperature, moisture and wind that meet along an irregular line variously referred to as a "discontinuity line," "polar front," "wind shift," etc. These masses of air, cold and dry from polar regions, warm and humid from equatorial, do not readily mix but tend to preserve their individual identities, the warm, moist air being forced to rise above and flow over the denser cold air, with resulting condensation and precipitation and other attendant phenomena which give us most of the stormy weather characteristics of temperate latitudes.[5]

He was basically explaining the role of weather fronts—for which terminology had yet to be settled—in the day-to-day weather within the middle latitudes. In the same article, he also stated that:

> there is nothing particularly new in the "Air Mass Analysis" concept of forecasting. It has been quite fully understood for many years, but its effective application has not been possible, because it requires a greater wealth of observational material than has heretofore been available.

One can almost detect an ounce of irritation coming from Gregg as if to say "Yes, we know how this works, but no, we couldn't have done that before because we did not have the necessary resources." In truth, because not enough observations existed (the number of surface observations was adequate but not optimal, and upper-observations were almost nonexistent until recently), forecasters could not explicitly apply this claimed knowledge. However, some of the most successful forecasters did instinctively use elements of this type of analysis. One of them was Charles L. Mitchell, the forecaster in charge when the Great New England Hurricane came under the responsibility of the Weather Bureau D.C. Office.[6]

The airmass analysis methods mentioned so many times in Weather Bureau reports and communications during the 1930s and 1940s involved much more than just detecting existing air masses. As applied to forecasting, the methods also required an understanding of the structure and behavior of the distinct bodies of air, which would in turn lead to more accurate forecasting of how the weather at a given location would be affected by air masses and their fronts. To do this, large-scale weather patterns must be observed not just at the surface but also in the air above. The needed analyses are equivalent or at least the precursors to the "synoptic analysis" methods of today. The word *synoptic*, as in "forming a synopsis," means all-encompassing or comprehensive and synoptic meteorology deals with the study of weather patterns and features, as can be summarized and seen in a map of a broad area (e.g., the continental United States).[7,8] To accomplish such analyses, forecasters must have adequate weather observation coverage, with high enough temporal and spatial resolution. The needed network of surface observations was perhaps still not fully suitable in 1938 to detect all the relevant weather patterns, but it had been rapidly growing throughout the previous five or six decades and would continue to do so. On the other hand, until then it had not been possible or even attempted to obtain reliable, systematic, and coordinated upper-air measurements. It is clear, then, that one of the most important advances following the need for the application of airmass analysis techniques was an upper-air network solution. It would take time to build, but once in place the resulting data would allow not only improvements in general forecasts but the ability to determine the steering currents at play when a hurricane was threatening land.

For this and many other reasons, the late 1930s and early 1940s were truly a transition time for the Weather Bureau. The death of Chief Gregg in mid-September of 1938 (see Chapter 1) occurred when the recommendations of the SAB were, though slowly, finally translating into reality via various initiatives:

- Surface observations were beginning to improve with better and more standardized instrumentation and more adequate locations for observation stations.
- Coordinated daily upper-air observations (to about 15,000 ft) were being obtained mostly by using instrumented planes through cooperation with the U.S. Army and Navy.
- Weather balloon technology was now becoming more feasible thanks to the use of radio transmission instead of relying in the physical recovery of the instruments.
- Forecasters were starting to be trained in airmass analysis techniques at the Central Office where the Air Mass Analysis section had been established.

- A test for new forecasters and those seeking promotion was developed.
- An initiative to send select forecasters to study at academic institutions was started.

It seems fair to say, therefore, that even with limited resources, the Weather Bureau did in good faith try to make as many improvements and adjustments as possible to follow the recommendations; but it is unrealistic in hindsight to expect that these changes would happen overnight, especially when breaking free from the status quo. The combination of the new Chief Francis W. Reichelderfer (see Chapter 1), the onset of World War II, and all the new technology and knowledge meant that the agency emerged from this period with no choice but to leap forward, landing in the age of weather radar, computers, and eventually, meteorological satellites. Into this time of historical transition at the Weather Bureau blew the 1938 Hurricane.

A History of Hurricane Forecasting

Historically, forecasting the behavior of a hurricane has been challenging, to say the least, especially before many of the tools on which we rely today became available. The unexpected arrival of hurricanes happened more often than not, sometimes seemingly materializing out of the blue. In fact, it was originally believed that all storms formed spontaneously in place.

Signs of a Hurricane

Even before much was known about the structure or behavior of hurricanes, many individuals from farmers to mariners whose livelihoods or very lives depended on weather began to recognize the clues that signified the imminent arrival of these extraordinarily devastating storms. Many early hurricane articles in the *Monthly Weather Review* (MWR) from the late 1800s to the early 1900s focus on describing the specific signs observed before a reported hurricane.[9] High cirrus clouds coming from the south (or any other odd direction for a specific region) were recognized as a possible sign of a nearby hurricane, though certainly not a guarantee. One of the most commonly noticed predictors was an ocean swell raising coastal water levels to a foot or two above normal, known to arrive anytime from one to three days before an incoming hurricane.[10] A strangely tinted sunset or sunrise, often described as "fire colored" or "brick-dust sky," was also commonly reported.[11] Of course, for those paying attention, more direct meteorological signs could also be detected as the storms were closing in. It was common to experience thick, humid conditions accompanied by higher than normal barometric pressure, followed by a gradual decrease (and finally followed by a sharper decrease, although at this time it was too late for any advanced warning). Much of this—call it art or wisdom—arguably might have been lost once people started to become reliant on an organized weather service that warned them of incoming storms. By the 1930s, most regular citizens were not paying attention to subtle signs in the sky, but a few still did. It has been reported, for example, that Portuguese fishermen in Connecticut, wary of a "copper-colored hue of the sunset" the previous evening, decided against going out on the morning of September 21, 1938.[12]

The first hurricane prediction in recorded history was made by Christopher Columbus during his fourth voyage to the New World in 1502.[13] During his first three expeditions, he learned much about tropical weather and the ominous signs of the exceptionally dangerous storms often encountered around the so-called West Indies—much different from the ordinary storms found over the eastern Atlantic and the Mediterranean Sea. During an unauthorized stop at the Santo Domingo harbor in what today is the Dominican Republic, Columbus sensed an imminent storm "from the hazy appearance of the atmosphere, the direction of high-flying cirrus and the presence of an ominous southeasterly swell." He sent a message to the governor of La Española (the Spanish name of the island), who had become his rival for the favor of the Spanish crown, urging him to keep his Spain-bound fleet of 30 ships safe in harbor until after the storm, but his warning fell on deaf ears. Although denied the protection of the harbor, he used his experience to head toward the lee side of the island (relative to the expected tempest) and thus found his way to sheltered waters. The storm did arrive two days later and at least 20 of the ships and about 500 sailors, who had begun their trip and were exposed in open waters, were lost.[14]

Early Understanding of the Nature of Storms

Much more knowledge about the behavior of both the atmosphere and storms in general gained during the next four centuries would still be essential to even attempt a modern hurricane forecast. Clearly, understanding the general circulation of the atmosphere that so much determines the basic track of tropical cyclones (see Chapter 4) is of utmost importance. The foundations for this knowledge did not come until the late 1600s and early 1700s. Before any of that knowledge could be used for a rough determination of the motion of hurricanes, however, it first had to be recognized that storms (not just hurricanes, but all storms) moved within the atmosphere and didn't just form and dissipate in place.[15] Furthermore, even after storms were recognized as moving systems, it was assumed that a storm came from the direction in which the storm winds blew; that is, until famous American founding father, inventor, and scientist Benjamin Franklin noticed something unexpected in 1743. A recent northeast storm (with winds blowing from the northeast), which turns out was actually a hurricane or at least its remnants, had prevented those in Philadelphia (including himself) from observing an anticipated total eclipse of the moon. The same storm was experienced in Boston (to the northeast of Philadelphia) later rather than earlier, and Bostonians were able to enjoy the eclipse before the storm commenced, meaning that the storm moved in the opposite direction than the observed winds. To Franklin, there was only one explanation: storms "were entities of atmospheric disturbance that move independently across the face of the planet."[16] This may seem like an elementary observation today, but at the time it was an essential insight upon which knowledge of the behavior of the atmosphere would subsequently build. It was the aggregation of these and other such insights that would ultimately lead to the formal advent of the science of meteorology and an understanding of storms in general, including hurricanes. Indeed, everything from the philosophy of scientific pursuit to the circumstances that forced mariners, weather enthusiasts, and scientists to ponder the nature of storms was at one point or another involved in the journey to understand the behavior of hurricanes, and thus, the ability to forecast them and warn the public of their possible arrival. The American Storm Controversy, previously described in Chapter 3 and resulting from William Redfield's observations of "southeast storms" or hurricanes being "whirlwinds" or rotating systems, is a perfect example of the relationship between science philosophy and the evolution of our knowledge about hurricanes. By the second half of the 19th century, it was finally understood that hurricanes are rotating storms forming in the tropics that move within the atmosphere.

Hurricane Forecasting in the 1800s and Early 1900s

Early hurricane forecasting relied in great part on actually detecting the presence of the storm, and then warning others of its potential imminent arrival. Advances during the late 1800s and early 1900s mostly involved devising and fine-tuning strategies to improve the ability to detect the storms with very limited data and communications. The establishment of observation networks and early tools (such as the telegraph and, later, the invention of radio communications) aided immensely in this endeavor. A pioneer of such hurricane forecasting was Padre Benito Viñes, the director of Belén College

"SIGNALING" WEATHER

There are various types of storm warnings in existence today for different types of severe weather such as tornadoes, severe thunderstorms, blizzards, and, of course hurricanes. Each one has explicit guidelines for its usage, but regardless of the specifics, they are all announcements to the public by a specific weather office or agency.* In the late 19th and early 20th centuries, all warnings were issued by the Weather Bureau and their "ordering" had a more physical meaning. A flag or specific configuration of flags was hoisted and displayed in specifically designated coastal signaling stations, mostly for the benefit of mariners and local residents. When the national weather observing and warning service was first organized in 1870, it was under the jurisdiction of the U.S. Army Signal Corps, which had dealt with the dissemination of information, often by using signaling flags, for the benefit of military operations since its establishment in 1860.† Originally named the "Division of Telegraphs and Reports for the Benefit of Commerce and Agriculture," the new service added weather observations and reporting to the original signaling duties—a welcome addition as such duties had severely dwindled since the end of the Civil War.‡,§ Shortly after the establishment of the weather service, special flags were devised for weather signaling. A

at Havana, Cuba. He used meticulous daily observations of ocean swells, winds, and high cloud motions, among others, to determine the specific patterns associated with the coming of a hurricane. He then used the knowledge he gained from those observations to develop the first hurricane forecasting and warning scheme in the 1870s.[17]

In 1938, hurricane forecasting was still relatively rudimentary, and in large part meant figuring out where the storm currently was, how fast it was moving, and in which direction. At this time the Weather Bureau had a good network for obtaining surface observations, including radio observations from ships at sea and a few useful land stations. It had limited knowledge of steering flow patterns via an inadequate network for obtaining upper-air observations. Very few of these observations were made over the East Coast, but they were supplemented by indirect observations of upper-air flow made possible by high clouds. Forecasters used a general knowledge of the current characteristics of surface features, such as the position of the Bermuda High, red flag with a black square center, for example, was a "cautionary signal" indicating a storm in the vicinity (originally only used in the Great Lakes and the eastern seaboard). Several flags were quickly added until a full set of weather signal flags was in place by 1885.**

---

* For example, the National Hurricane Center issues hurricane watches and warnings, the Storm Prediction Center issues severe thunderstorm and tornado watches, and individual National Weather Service Forecast Offices issue severe thunderstorm and tornado warnings in their locally designated areas.

† The U.S. Army Signal Corps' ongoing mission is to provide and manage communications and information systems support for the command and control of combined armed forces (as stated on their website).

‡ See, e.g., MWR (July 1872), which summarized storms, rainfall, and temperatures observed during the month throughout the United States. The header reads: "Monthly Weather Review, July, 1972. War Department, Office of the Chief Signal Officer, Division of Telegrams and Reports for the Benefit of Commerce and Agriculture." This was just a one-page report that eventually grew into the peer-reviewed meteorological research journal that it is today.

§ For more information on the history of the establishment of the U.S. Army Signal Corps see Raines (1996) and for more on the early history of the U.S. weather services see Whitnah (1961).

** From various Annual Reports of the Chief Signal Officer from 1871 to 1885, as can be found in Record Group 111 at the National Archives and as detailed in Raines (1996), Chapter 2 endnotes.

and guesswork about the hurricane's progress over the ensuing 12 hours, much of it based on previous experience (in other words, persistence and climatology). It might sound like a simple or inaccurate approach, but it was in fact an impressive achievement resulting from decades (one could say even centuries) of accumulated knowledge about hurricanes and their behavior.

The Weather Bureau had done such a good job of forecasting hurricanes over the previous couple of decades, especially for storms in the Gulf of Mexico, that accurate forecasting or at least warning was expected by the public. However, because hurricanes were a relatively rare phenomenon in the Northeast (even unheard of, to many), that portion of the public would not have expected such storm warnings on a regular basis, if ever.

Recognizing the Signs of the 1938 Hurricane

Besides the Weather Bureau, there were others keeping an eye on the storm. Amateur and professional meteorologists from universities and observatories and even some independent forecasters, after hearing about a hurricane offshore for the previous few days, had been paying attention to the weather reports and observing the clouds and the barometer. John Q. Stewart, astronomy professor at Princeton University and amateur meteorologist, wrote an "explanation of the storm" for *Harper's* magazine, where he described having noticed the hazy, humid conditions and cloud motions while perched on the White Mountains of New Hampshire. A local personality self-nicknamed the "Old Sail," who wrote a weather column for the *Gloucester Daily Times*, had also been monitoring the advisories and general meteorological conditions since the storm was approaching Florida. During the morning of the day of the Hurricane he declared that "the tropical hurricane . . . is expected to cross Southern New England between sunset and midnight tonight" as part of his published forecast.[18] While the Washington, D.C., Office in charge of the official forecast was apparently still clueless, this obscure local was right on the money, though his forecasted timing of the storm was still too slow.

The observer at Blue Hill Observatory in Massachusetts speculated early in that fateful week that the tropical air in the area could mean that the storm would not be able to recurve.

> Sept. 18. When the hurricane was first reported . . . the fact was noted . . . that since the eastern United States was covered by a considerable flow of moist tropical air the hurricane could not be expected to recurve at once and so would probably reach the coast.

Even more, throughout the morning of the 21st, he consistently noticed south and southeast winds in the upper atmosphere by observing the direction of motion of high clouds observable at times through breaks in the lower clouds. His morning forecast read as follows:

> Cloud observations indicate that the whole depth of the atmosphere up to at least ten km is moving from a little east of south and therefore the storm south of New England must move inland. Since the airwaves broadcasts show very low pressure at Lake Hurst [New Jersey] and Michel Field [Long Island], the center will probably move northward or a little west of north through western New England or possibly eastern New York and at high velocity. There should be intense rainfall but not for long. The center should be well past and the weather clearing with winds shifted to SW by midnight.[19]

Tracking the 1938 Hurricane North of Cape Hatteras

Starting the day before the Hurricane's landfall, while Jacksonville was still in charge of its tracking and forecasting, the Washington, D.C., Office ordered northeast storm warnings for areas under its jurisdiction north of Cape Hatteras:

> Advisory one P.M. [September 20] Northeast storm warnings ordered south of Virginia Capes to Cape Hatteras. Wind will become north or northeast this afternoon and probably increase to gale force late tonight or Wednesday forenoon.[20]

In other words, D.C. clearly expected strong northeast winds—probably increasing to tropical storm strength—in coastal areas (with the storm being offshore) between Cape Hatteras and Virginia on Tuesday night or Wednesday morning.

A northeast storm warning, such as the ones ordered on September 20 and 21, consisted of a red pennant flag above a square red flag with a small black square in the center. A hurricane warning, which was never ordered for this storm, was represented by two of the same red square flags with black centers arranged vertically. For more than 100 years, marine signals using combinations of flags during the day and lights (first lanterns and later electric lights) during the night were displayed at specially designated stations managed by the Weather Bureau. It also made sense for many of these coastal locations to be colocated with Coast Guard stations, marinas, and other visible coastal sites. The Weather Bureau would call each station individually to order the raising or dropping of the signals, a task that eventually became too resource-intensive and somewhat obsolete because of continuous radio broadcasts of forecasts and warnings via the NOAA Weather Radio.[21] In 1989, the National Weather Service discontinued this National Coastal Warning Display Network, seemingly putting an end to the era of weather signaling. In 2007, however, the U.S. Coast Guard reestablished a limited coastal warning display program at selected stations mainly serving small craft, with the rationale that the system could still save lives. The suite of weather signal flags in use today is more limited than the original set from the late 1800s. Northeast and southeast warnings (referring to wind direction) are not specified anymore. Hurricane warning flags, on the other hand, are exactly the same as they were since the beginning of the program. Additionally, the modern flags used for small craft advisories

Warning flags were hoisted on signal towers at display stations such as this one at Delaware Breakwater (NOAA Photo Library).

(one red pennant) and gale warnings (two red pennants) were not included as part of the original signaling roster. The current system, updated in 1958, emphasizes the strength of the storm systems rather than the wind direction expected with their arrival.[22]

At 9:30 P.M. on September 20, the warnings were extended to Atlantic City, with a brief mention of a "tropical storm" (referring to the storm's ori-

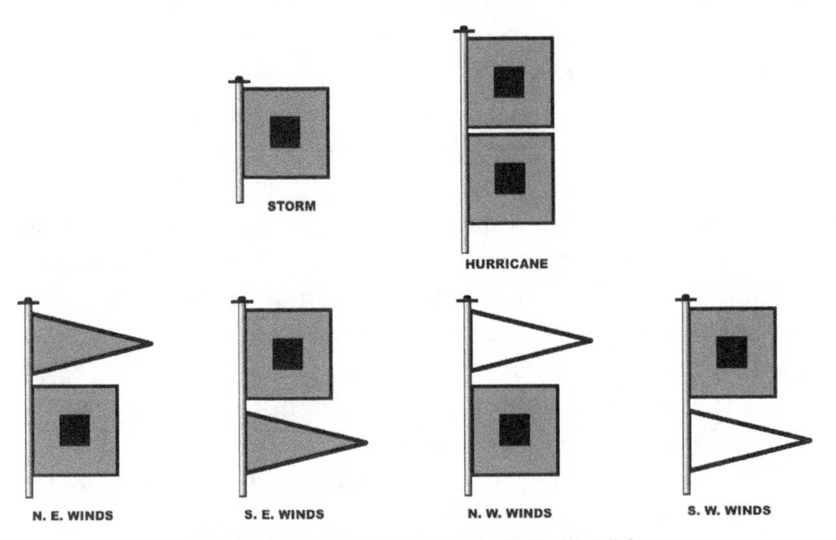

Weather signal flags used at Weather Bureau Display Stations during the late 1800s and early 1900s (adapted from Abbe [1899], *The Aims and Methods of Meteorological Work*).

gin rather than its strength) expected to bring gale force winds to the New Jersey coast late Wednesday. The next morning (September 21), after official responsibility was handed off to D.C. (which had already been taking care of warnings north of Cape Hatteras), the advisories continued, but unlike the detailed informational statements that came out of Jacksonville earlier, these contained mostly updates of ordered warnings. Only a 9 A.M. advisory (which also extended storm warnings to Eastport, Maine) contained information about the storm itself:

> Advisory nine A.M. [September 21] Northeast storm warnings ordered north of Atlantic City and south of Block Island and southeast storm warnings ordered Block Island to Eastport Maine. Tropical Storm apparently central about seventy five miles east of Cape Hatteras moving rapidly north-northeastward attended by shifting gales over a wide area and by winds of hurricane force near its center. Northeast or north gales backing to northwest south of Block Island to Hatteras today and southeast to east gales Block Island to Eastport becoming northwest tonight or Thursday morning. Small craft should remain in port until storm passes.

It is clear that at 9 A.M. the storm was already known to be moving rapidly, although it was certainly still moving faster than projected, with Thursday morning mentioned as a possible time when it would be coming near enough to the Northeast to be felt. It was also known that the storm still possessed hurricane-strength winds near its center, but this was to be expected while it was still completing its recurvature process. It would also be expected that the storm should now start to weaken during its northward movement into a less favorable environment. The mention of north-north-eastward movement shows that Weather Bureau forecasters still anticipated that it would continue recurving. Southeast winds in Maine would be expected if the storm made landfall, but it was clear that was not expected. The only other way southeast winds could be experienced in this area, more likely the expectation in this case, would be well ahead and north of the storm and before it continued its recurvature onto the waters of the North Atlantic.

By late morning, there was some awareness that the storm was not decreasing in intensity as fast as expected, but the hurricane terminology was nevertheless abandoned, with forecasters obviously clinging to the notion of a weakening storm.

> Advisory eleven thirty A.M. [September 21] Warnings changed to whole gale Atlantic coast north of Virginia Capes to Sandy Hook New Jersey. Tropical storm cent[ered] ten A.M. about one hundred miles east of the Virginia Capes moving rapidly northward or slightly east of north. It is attended by shifting gales over a wide area and by winds of whole gale force over a considerable area around center. Northerly winds along the New Jersey, Maryland and southern Delaware coast will likely increase to whole gale force this afternoon and back to northwest and diminish tonight.

A whole gale corresponds to the strength of a moderate to strong tropical storm (55–63 mph), which means that the warnings were indeed upgraded in urgency, though not to the levels that a hurricane warning would have produced. The strong winds were, in fact, expanding in radius as the storm lost its tropical nature and became an extratropical cyclone (see Chapter 7 for details on the extratropical transition of the Hurricane). Once again, the "tropical storm" terminology most likely refers to the origin of the storm rather than to its specific strength. The storm was now known to be closer to the coast than before, so perhaps there could have been a suspicion that it was not recurving, but there is no sign in the advisory that such suspicion existed yet.

Shortly after the 11:30 A.M. advisory it became obvious that the storm was going to hit land. Just two and a half hours later at 2:00 P.M. one last advisory was transmitted, although it still did not specifically mention that the incoming storm was a hurricane.

> Advisory two P.M. [September 21] Warnings changed to northwest Virginia Capes to Sandy Hook New Jersey. Tropical storm cent[ered] twelve noon about seventy five miles east-southeast of Atlantic City moving rapidly north-northeastward with no material change in intensity since morning. Storm center will likely pass over Long Island and Connecticut late this afternoon or early tonight attended by shifting gales.

This is the first acknowledgment of the absence of recurvature, or rather, recurvature is not mentioned or implied anymore. It was clear now that Long Island and New England would be imminently hit; but even then, forecasters did not appear to suspect that the storm was still as strong or moving as fast as it was, as evidenced by the expectation of a likely "early tonight" landfall.

The advisories perhaps paint a somewhat simple picture of the knowledge of the storm just prior to landfall and what warnings were issued; but behind the scenes, the story was certainly much more intriguing.

The Not-So-Junior Forecaster

Stories about the Great New England Hurricane of 1938 told during the last two or three decades invariably mention a junior forecaster who was the only one to correctly predict, against his superior's opinion, that the storm would hit the Northeast. Charles H. Pierce, who indeed accurately predicted the storm's landfall by carefully analyzing the few available observations and correctly interpreting what those observations indicated about the track the storm would follow, was not as inexperienced a novice as might be inferred from these recent stories.

Although Charlie Pierce, as he was known to his colleagues, had only been hired at the Weather Bureau less than a year before (listed as a new hire for November 1937), he had something that many forecasters did not have at that time: a college degree with extensive training in state-of-the art meteorological analysis.[23] As a young student in Ohio during the late 1920s, he learned of the increasing demand for Weather Bureau meteorologists created by the aviation industry and decided to give the field a try. In 1930 he transferred to Clark University in Worcester, Massachusetts, and studied under Charles F. Brooks, who was one of the founders of the American

Meteorological Society and would soon become the director of the famous nearby Blue Hill Observatory.[24] He also spent some time at MIT, where renowned meteorologist Carl-Gustav Rossby (who brought with him the latest Norwegian analysis techniques) had recently started a program that combined the often disconnected theoretical study of the atmosphere and practical forecasting.[25] Finally, he received a Bachelor's degree from Boston University in 1933.[26] As an undergraduate student, Pierce worked at the Blue Hill Observatory as an observer and research assistant. Afterward, degree in hand, he worked as a Trans World Airlines (TWA) meteorologist for a few years before finally joining the Weather Bureau in 1937.

The way in which forecasters were trained was starting to change during the mid-1930s. The path taken by Pierce—somewhat similar to the path taken by many modern forecasters—was very different from the on-the-job-training by which most Weather Bureau forecasters had learned their trade. Experience was a valuable and valued asset to a weather forecaster, but now many young men (and it was all men before World War II) were starting to have the opportunity to gain a formal education in meteorology.[27] This new breed of forecasters were trained in the state-of-the-art airmass analysis techniques that had been so much encouraged during recent years and recognized as essential to more accurate forecasting. This knowledge and proficiency on the new techniques put them on par with those who had to spend years gaining knowledge and intuition on the job. Additionally, upper-air observations were essential to flying and Pierce had come to the Weather Bureau directly from his aviation industry forecasting job. It stands to reason, then, that he came to the Bureau with more knowledge of frontal and upper-air processes and techniques than the average forecaster, especially for a Weather Bureau rookie.

The Professional and Technical Services of the Washington, D.C., Weather Bureau Central Office consisted of seven divisions in 1938: Instrument, Aerological, Climate and Crop Weather, River and Flood, Marine, Meteorological Research, and of course, Forecasting.[28,29] Most of the Meteorological Research Division was devoted to airmass analysis, which over the last four years had grown to a group of eight meteorologists, including Pierce, who was hired at the Assistant Meteorologist level (at an annual salary of $2,600, which would correspond to approximately $42,300 in 2012 USD).[30] At the time, there were only three senior forecasters in Washington, D.C. Charles L. Mitchell was considered the "official in charge," the equivalent of today's Meteorologist in Charge (MIC). He was also the senior forecaster on duty on the day of the Hurricane. The Forecasting division, which provided assistance to the senior forecasters via what they called the "map and radio force," was an army of 16 analysts who hand-plotted and analyzed maps.

Many of these young men, classified as junior or assistant meteorologists, occasionally produced practice forecasts to be supervised by their seniors and compared against the official forecast.[31] The airmass group, as it was known, did not generate any official forecasts, but mostly engaged in research and training activities.[32] They did, however, lead a midday weather discussion based on their map analyses.[33] There was no actual forecasting as part of Pierce's normal duties, but as a former aviation meteorologist, it is safe to assume that he was used to producing forecasts for which he had to pay close attention to upper-air observations (limited as they were at the time).

None of the early accounts of the storm, even those criticizing the actions of the Weather Bureau, mentioned anything about a forecaster within the ranks correctly forecasting the track of the storm against his superiors' opinion. Two publications coinciding with the 50th anniversary of the storm in 1988 also fail to mention this story.[34] It wasn't until a 1993 PBS television documentary about the storm that an unnamed junior forecaster was mentioned (incorrectly stating that he had drawn a map showing the correct storm movement a whole 24 hours in advance).[35] Brief obituary articles written after Pierce's death in 1994, however, describe him as "a forecaster for the National Weather Service [who] accurately predicted the path of the devastating hurricane of 1938." Then, in 2003, in her book *Sudden Sea*, Angela Scotti told a detailed account of the role of Pierce and the events at the Weather Bureau on that day in her book, and since then, the story has been treated as common knowledge, repeated by several authors with some slight variations in the details, all using Scotti's account as a source.[36] Based on the resources at her disposal, Scotti built a likely scenario describing what she understood of the events on that day. However, an original or primary source describing the actual events hasn't surfaced yet, meaning that there are portions of her story that cannot be corroborated or rejected at this time. On the other hand, some direct and indirect facts were indeed uncovered, with some of them verifying and some contradicting portions of the story as told by her. The following discussion is both based upon those facts and supplemented by Scotti's account where no other evidence was available.

Pierce was apparently pulled from the ranks to fill in for a forecaster on vacation that day, and he was tasked with tracking the storm that had been lurking over the western Atlantic waters and that had been monitored by Jacksonville until early that morning.[37] "The ranks" from which he was

Charles H. Pierce, circa 1980s, after a distinguished forecasting career in the Weather Bureau (American Meteorological Society).

pulled would have been those of the airmass analysis group. (A November 1938 Weather Bureau personnel roster shows Pierce as part of this group, which has led to the common assumption that he was promoted to that position as a reward for having correctly forecasted the storm. It is more likely, however, given the close timing of the roster to the storm and Pierce's previous experience and credentials, that he was hired directly into that group). He was also most likely asked to join the "map force" for the day to draw map analyses based on the available observations. That he was chosen among his peers might be a sign that his exceptional analysis skills were recognized by his superiors. Little did he know that his originally seemingly obscure role in the forecasting of this storm would lead, many years later, to the distinction of being the only forecaster to correctly determine the storm's path.

Before continuing the story of the events on September 21, it is important to note that perhaps of more significance, Pierce was the first author to show without a doubt that tropical cyclones can undergo an extratropical transition. He did this within the year following the storm by performing a very detailed analysis of the surface and upper-air observations and correctly interpreting their significance.[38] A review of much of the tropical cyclone–

related literature (mostly from MWR and supplements and from special Weather Bureau bulletins) from the late 1800s to the time of the Hurricane does not reveal evidence of any other article describing this transition, except for a brief opinion mentioned in a review note concerning typhoon tracks stating that "when sufficient pressure data are available a close study will prove that most typhoons give rise to what is generally known as extratropical cyclones."[39] It appears that Pierce provided this "close study," that would prove the occurrence of extratropical transition, though for the so-called West Indian hurricanes rather than the Pacific typhoons (both types of storm are, of course, tropical cyclones affecting different parts of the world).

Pierce's 1939 analysis of the "meteorological history" of the Hurricane is a remarkable piece of meteorological analysis. All descriptions and accounts of previous hurricanes consisted mostly of describing observations and effects of the different storms. Even summary supplements and bulletins focused on storm tracks and statistics, and the conclusions therein were based on such statistical observations.[40] Pierce, on the other hand, performed a detailed analysis of surface maps (which will be used in Chapter 6 as part of the description of the meteorology of the storm), including frontal boundaries and upper maps showing the steering of the storm. In the process, he used all of the state-of-the art airmass analysis techniques and all the theoretical knowledge at his disposal to interpret his results. Even more, his article is reminiscent of modern synoptic-scale case studies, with much more modern terminology and certainly more meteorological analysis than any previous hurricane study had included before.

"Hurricanes Avoid New England"

As Pierce closely analyzed the data at hand, he must certainly have noticed clues that the official thinking that the Hurricane was recurving (as first posted by Jacksonville and now adopted by Washington, D.C.) was not consistent with the evidence. Very little upper-air data were available on the spot, but there was at least one station report, it turns out from Washington, D.C., indicating southerly winds. Assuming that such flow was characteristic of the upper conditions throughout the region and not just above D.C., it would have been easy to conclude that the upper currents would steer the Hurricane northward and not northeastward.

Scotti describes a noon forecast meeting where Pierce eagerly presented his maps and evidence at odds with the official forecast. This so-called meeting was most likely the daily weather briefing by the airmass analysis section, as described in the 1938 Weather Bureau Chief Report.

> This small group, usually referred to as the "air-mass group" prepares daily a general map, vertical cross sections [and other specialized charts, as needed]. Conferences are held at 11:15 A.M. at which a leader carries through a complete analysis of the current map and traces the evolution of the fronts and the changes in air masses. . . . These conferences serve not only to instruct personnel who are engaged . . . but assist the forecaster on duty by providing a complete and thorough analysis of the morning chart for the current day. This analysis is of considerable value even though it follows the making of the morning forecast.[41]

This description was written by Chief Gregg shortly before his death, which as mentioned earlier, occurred just a week before the storm. It is possible that Pierce was indeed the presenter or at least contributed to the discussion. However, rather than a forecast meeting where the official forecast and warnings were discussed, this "conference" was apparently an educational meeting for any interested Bureau personnel, including forecasters from the so-called field offices who often came to the Central Office to become versed on airmass analysis techniques. It is clear that the forecaster on duty would have been in complete control of the official forecast, and it is implied that he might or might not use the information learned from these analyses for his later forecast, depending on how valuable they seemed in each individual case.

As the story continues, Pierce was unequivocally and unceremoniously told that severe hurricanes did not affect New England. Scotti also mentions a senior forecaster declaring that "only a 100-year storm" could follow that path and severely affect the region. It is unclear who this senior forecaster might have been but, given the frequency of such severe hurricanes capable of producing widespread devastation throughout the entire region, he would have actually been correct. A 100-year storm means that in any given year there is only a 1 in 100, or 1%, chance of such a storm occurring. This is the kind of storm that the 1938 Hurricane turned out to be. In fact, the probability for such a storm is actually lower than 1% for most of the region (see Chapter 9 for more information about the frequency of New England hurricanes). No major hurricane had affected New England since the establishment of the Weather Bureau, even dating back to the early military years of the agency in the 1870s, and conventional wisdom was, with good reason, that hurricanes "avoid New England." Forecasters knew how the storms tended to recurve and weaken as they steered around the Bermuda High and reached colder waters. They also knew that those storms that did come to the region did so in a weakened state. It would have seemed, therefore, that there was no reason for alarm.

The most recent accounts of the storm paint a picture where the senior forecaster on duty, Charles Mitchell, is a sort of villain who arrogantly ignored the undeniable evidence being presented to him and overruled the forecast and warning that would have saved hundreds of lives. The events on that day were most likely not as black-and-white.

Mitchell was celebrated as one of the best forecasters in the nation. He had joined the Weather Bureau more than 30 years before and, thanks to his awe-inspiring, almost supernatural forecasting skills, had become a senior forecaster for the top forecasting district in the nation (Washington, D.C.) in 1925.[42] He has been described as having an uncanny ability to "astutely mix generous doses of human judgment with objectivity in formulating his predictions . . . almost as though [he] had an analogue computer in his brain." Recognizing the importance of continuity, he and his fellow D.C. senior forecasters would have grueling month-long shifts every other month, producing and overseeing every forecast for seven days a week and being on-call for 24 hours a day. During his month off, after resting for a few days, he would then focus on research activities and on studying normal weather for his upcoming month-long shift. His extraordinary instinct was accompanied by careful study of all observations at his disposal, incorporating any new available information and techniques. He was recognized by his peers, pupils, and successors during his lifetime, and posthumously right after his death in 1970, as the most eminent forecaster of the first half of the 20th century. He was also a champion for continuing the program of upper-air observations beyond an initial research period, as he foresaw its potential usefulness to forecasting. That said, Mitchell was quite a character. He knew he was good at his job, and he joked dismissively about new "fad" techniques (some of which he actually used directly or indirectly in formulating his own forecasts). He also expected nothing less than excellence from those around him, often frustrating the map plotters by finding errors or inconsistencies in the maps within seconds. But according to those remembering him long after his death, he had great likeability and was a "gentleman of the old school in the finest sense of that phrase and most who worked with him treasure the opportunity they had not only to profit from his knowledge, but also to spend quality time with such a fine person."[43]

Mitchell was also the author of a study titled "West Indian Hurricanes" in 1924, which he completed while at the Bureau's Marine Division. He performed extensive analysis of all the data available on Atlantic tropical

Eastport to Sandy Hook, fresh southwest winds and fair weather to-night and Friday.

Sandy Hook to Hatteras, fresh southwest or west winds over extreme north portion and moderate southwest or west winds over central and south portions and fair weather to-night and Friday.

WEATHER CONDITIONS LAST 24 HOURS

The storm of tropical origin that was central a short distance east of Cape Hatteras Wednesday morning moved very rapidly northward to Vermont and from there northwestward to southwestern Quebec with diminishing intensity, Doucet, 29.16 inches. The center of this storm passed inland with undiminished intensity near New Haven, Conn., at 3:40 p.m., Wednesday when the barometer fell to 28.10 inches. The storm was attended by heavy rains in New England, New York, and portions of the Middle Atlantic States, and by strong shifting gales over a wide area. The greatest 24-hour rainfall reported was 3.44 inches at Atlantic City, N. J., and the highest reported wind velocity was 88 miles per hour from the southeast at Boston, Mass. Pressure remains low from Alaska eastward to Iceland, Reykjavik, Iceland, 29.15 inches; Kodiak, Alaska, 29.26 inches, and Chesterfield, District of Keewatin, 29.28 inches. Pressure is high over British Columbia and from Nevada and southern Utah east-southeastward to the Gulf and South Atlantic States, and over the ocean east and northeast of Bermuda, Kamloops, B. C., and Durango, Colo., 30.16 inches; Shreveport, La., and Cape Race, N. F., 30.12 inches. The temperature has fallen in the Atlantic States, while it has risen in the Plains States, the upper Mississippi Valley, and the upper Lake region.

C. L. MITCHELL.

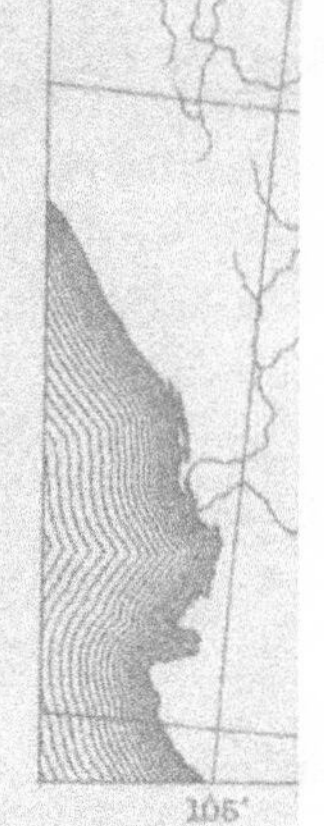

Excerpt from the U.S. Department of Agriculture Weather Bureau Daily Weather Map for September 22, 1938, showing a brief description of the Hurricane signed by C. L. Mitchell as part of the "Weather Conditions Last 24 hours" (NOAA Central Library).

cyclones since the late 1800s. The work was the most complete of its kind at the time and for several decades afterward. As part of the study, he examined the effects of anticyclones, or areas of high pressure, on the tracks of tropical cyclones. He stated that:

> it is well understood that no tropical cyclone will recurve in the Atlantic Ocean or the Caribbean Sea so long as the more or less permanent anticyclone that extends from the vicinity of the Azores west-southwest-ward over Bermuda to the coast of the United States persists; it will be carried along in the general drift of the atmosphere at the higher levels, say from 3 to 5 kilometers above the surface, and it will skirt the southern end of this anticyclone.[44]

As will be discussed in the next chapter, this was exactly the situation at hand, with a strong Bermuda High and high-level steering currents that would take the storm northward and prevent the storm from recurving northeastward toward the North Atlantic. It is perplexing that Mitchell would ignore what he himself had concluded and clearly understood. A handful of contemporary critics who knew him personally indeed claimed that the imminent track of the storm should have been obvious from the available observations, even if sparse, and especially to Mitchell.[45]

As the story continues, Mitchell was not to be second-guessed and made sure that the word *hurricane* did not appear in any of the official public advisories coming out of the office. It is true that most of the advisories after

Washington, D.C., took over do not mention a hurricane. Even the next day's Daily Weather Map summary signed by Mitchell did not specifically mention a hurricane, calling it instead a storm "of tropical origin."

It seems, however, that the more likely scenario is that from the perspective of the monitoring and forecasting by the Washington, D.C., Office, the storm didn't seem any different from various other storms of tropical origin and of similar track over recent years, all of which had either recurved northeastward (as this one seemed to be doing) or weakened significantly (as they reached colder waters and a less favorable environment, especially the enhanced wind shear within the higher latitudes). There was not enough data to clearly warn of the exceptional conditions that would not only bring this storm over New England but also cause it to move much faster than had ever been observed for such a storm and keep it from weakening as quickly as expected. All signs point to the fact that Pierce, indeed, noticed enough evidence and understood what was going on (being such a skillful synoptician, as evidenced by his soon-to-be-published groundbreaking analysis paper on the storm). However, Pierce's lack of rank meant that his forecast would not have carried much weight and Mitchell's experience would have trumped any alternative scenario presented by Pierce. His forecast, at least the written version, was very likely one of those practice forecasts done by assistant meteorologists and meant for later verification.[46] The midday briefing was never a means for advising the lead forecaster but rather a showcase of the day's weather, which due to timing issues regarding newspapers and broadcasting deadlines would happen after the lead forecaster had already made his day's forecast at 9 A.M.[47] Moreover, it might not be outrageous to say that most of the time Mitchell would have been the one who was right and the inexperienced junior forecaster or analyst would be the one learning from his experience; but this one time at least, when it mattered the most, Mitchell was wrong or seemingly unaware of what was going on in the atmosphere.

Both men went on to have long careers within the Weather Bureau. In modern accounts Mitchell is incorrectly reported to have quietly left the Bureau, but that was not the case. He continued as the most senior forecaster in Washington, D.C., for years and did not retire until 1950.[48] His month-long shifts continued (as evidenced by his signature on hundreds of Daily Weather Maps) until World War II, when he was recruited to provide long-range forecasts (seven days in advance, which would have been unheard of at the time) in support of military operations.[49] He received various awards for his forecasting contributions. Years after his death, an American

Meteorological Society award "in recognition of long-term service by individuals engaged in weather forecasting activities" was named in his honor. To this day, the Charles L. Mitchell Award is the Award for Outstanding Service by a Weather Forecaster, which is "presented to individuals who, through performance of exceptional forecast service, have distinguished themselves and brought credit to the profession."[50] Pierce, later "known as one of the premier forecasters in the country," continued with the Bureau until his retirement in 1974. Most of his forecasting career was spent in Boston, where he was first part of the founding crew of the forecasting office at Logan Airport, and later served as supervisor of the Guidance Forecasters. In 1970 he was awarded a Department of Commerce Bronze Medal for "valuable contributions to scientific weather forecasting," and in 1972 he received a National Weather Service Professionalism award "for an outstanding contribution to the advancement of applied meteorology."[51] It is fair to say that both men represented the best of the "old ways" and "new ways," respectively.

Warning the Northeast

The Weather Bureau quickly endured a barrage of public criticism following the storm, especially from local newspaper editorials.[52] The agency defended its actions, claiming that its forecasters did issue several storm warnings that undoubtedly saved lives and that their advisories contained as much information about the Hurricane as was possible under the circumstances. As mentioned earlier, a handful of educated amateur and professional meteorologists in the region also openly criticized the Bureau. They claimed that the warnings were not any different from those issued dozens of times for winter storms every year and that a rudimentary look at the winds above Washington, D.C., and the position of the Bermuda High should have told them everything they needed to know about the storm's path.[53] They also emphasized how most of the public did not hear warnings of any kind.

The truth is, once again, more complicated than either side claims. The Weather Bureau did issue storm warnings, most of them "northeast storm" warnings, and at some point the warnings for at least north of Virginia to New Jersey were updated to "whole gale," which would have carried more urgency. Also, detailed information about the whereabouts of the Hurricane was indeed put into many of the advisories. However, what the official post-storm statements, reports, and explanatory letters did not emphasize was that most of the detailed hurricane information came from the Jacksonville Office while the Northeast was not yet threatened. After responsibility was handed off to D.C., the word *hurricane* essentially disappeared from the

The cover art for the commemorative collection of photographs titled "It DID Happen Here!" is a pen-and-ink drawing by Tyleston R. Minor, showing West Street in Keene, New Hampshire, with a portion of the public library depicted to the right.

advisories and no actual hurricane warnings were ever issued. Would they have been issued even if forecasters knew the correct path of the storm? We will never know for sure. The assumption would have been that the storm was weakening, and they did not have much of a way to know how fast the storm was, in fact, moving. Thinking that they still had time (since hurricane warnings were normally ordered when the storm was believed imminent), the forecasters would most likely have wanted to be cautious so as not to cause panic to such a highly populated area before they were sure of the danger.[54] From this side of history, it's easy to oversimplify the situation. With such limited data, it would have been easily possible for either Pierce or Mitchell to be wrong.

As if responding to the common early-20th-century notion that intense hurricanes did not happen in New England, a commemorative photographic collection of the damages in southern New Hampshire published shortly after the storm answers emphatically with its title: "It DID Happen Here!"[55]

Decades later, in *Hurricane!*, his now out-of-print account of the storm, author Joe McCarthy clearly stated:

> [T]he unexpected storm disaster of 1938 did . . . perform one invaluable service, . . . to change the concepts of hurricane forecasting on the Atlantic coast . . . [and destroy] for all time the dangerous assumption accepted so complacently before then, that big tropical cyclonic storms were not likely to threaten the heavily populated shores north of Cape Hatteras.[56]

And so it was that on September 21, 1938, unknown to forecasters and residents, a deadly and exceptionally devastating storm was indeed racing toward the populated shores of Long Island and New England.

6

SUDDEN DEVASTATION

Unexpected landfall of an intense hurricane and what made it particularly devastating

Early in the morning of September 21, 1938, as the Jacksonville, Florida, Weather Bureau Office wrapped up its duties and the Washington, D.C., Office took over, the Hurricane was still far from the Northeast coast. At this time, with few remaining ships in its path, the storm stealthily raced offshore, advancing faster than anyone thought possible. In just a few hours, Long Island and southern New England would be hit head-on and with great force. For such a strong storm to slam into Long Island and New England, a series of particular meteorological conditions had to be in place.

Meteorology of a New England Hurricane

Stormy lows and calm highs commonly depicted on middle-latitude surface weather maps are a manifestation of the weather patterns aloft that can support or impair them.[1] Upper-air divergence (air vacating a column of air, thus counteracting surface inflow caused by the presence of the low itself) is needed for the formation, maintenance, or strengthening of centers of low pressure at the surface.[2] Upper-air convergence (air adding into a column of air, thus counteracting surface outflow caused by the high) is equivalently associated with surface highs. The way in which this happens involves a balancing act by the same atmospheric forces already described in Chapter 4: the pressure gradient, Coriolis, and centrifugal forces. Away from the influence of the ground's friction, the resulting air flow is along the constant pressure contours, or isobars, drawn in constant-height maps.[3] (Likewise, in a constant-pressure map, which is more commonly used today, the flow is along the contours of constant height.) The motion of air above the surface can thus be determined from upper maps by simply examining the isobars. The air accelerates and slows down as it moves along these pressure patterns, either spreading out or squeezing together, thereby producing the needed upper-air divergence to support surface lows (or the upper-air convergence to support surface highs).[4] These supporting patterns can be found along the axis of the jet stream, a narrow atmospheric river where the swiftest winds in the upper troposphere flow.[5]

Whether within the middle latitudes, the tropics, or otherwise, weather systems in the atmosphere move within their background flow, which meteorologists in the early 20th century often called the "drift of the atmosphere." Thus steered, hurricanes often leave the tropical latitudes in which they originate and reach the middle latitudes, where they of course continue moving within the general flow they encounter. The middle-latitude background flow contains the very changeable jet stream, with its associated troughs and ridges.[6] The specific timing of the arrival of a hurricane will determine with what features it will interact, and the results of the interaction can have sharply contrasting effects on the track of a storm, its development, and its ultimate fate. A trough dipping over the eastern United States will steer a hurricane in a general northward direction (in the same direction that air flows through the east side of the trough). A ridge over the eastern United States, on the other hand, will not allow a hurricane to turn northward along or near the east coast.

In the deep tropics, the steering currents are generally weak, and as described in Chapter 4, the early track of the 1938 Hurricane would have resulted from the easterly trade winds in combination with the clockwise flow around the Bermuda High. These large-scale atmospheric circulation features, together with a natural tendency for storms at low latitudes to drift northwestward, are what move the storms toward the subtropical and middle latitudes.[7] The specific way in which this happens depends on the strength and location of the subtropical Bermuda High as well as the shape, strength, and timing of the upper nontropical troughs and ridges as the hurricane approaches the western half of the Atlantic. The result is a large variability in possible track features such as the speed at which a hurricane moves and the longitude of its recurvature, if it recurves at all.

As the Great Hurricane approached the East Coast of the United States in September 1938, the North Atlantic was dominated by a well-defined upper ridge that provided additional upper convergence of air to increase the surface pressure over the Atlantic. The normally occurring Bermuda High was thus enhanced. It was stronger, centered farther west, and reached into higher latitudes than normal. Additionally, a well-established upper-level trough was settling over the eastern half of the United States. The flow associated with the large Bermuda High kept the Hurricane moving westward well into the western Atlantic Ocean. As the storm then came closer to the U.S. coast, and as it was starting to recurve, the trough initially provided a southwest flow. This flow would have steered the storm northeastward, as expected. However, the strong and large Bermuda High would not allow northeastward recurvature to continue into the North Atlantic. At the same time, the trough over the eastern United States was becoming stronger and reaching farther south. It extended all the way from Canada to the southeastern United States, and by the time the Hurricane progressed beyond subtropical latitudes, the steering flow had become more northward than northeastward. Troughs extending so far south into the United States tend to be more common during winter and late fall. The Atlantic hurricane season, on the other hand, spans from summer to early fall, with August and September being most active. Consequently, hurricanes and such deep troughs do not commonly interact.

This hurricane would thus move northward and never continue its recurvature. Even with the limited upper-air data available to him, Charlie Pierce was able to reconstruct much of this scenario in his "Meteorological History of the Storm" (see his upper-air maps in the next section). With much more data now available, the meteorological patterns accompanying a small number of hurricanes that reached New England since then become easily apparent (even though they were all much less intense and destructive than the 1938 storm). A very informative analysis technique is to create composite maps by averaging the conditions for each of the storms. The composite maps show the same meteorological features discussed above: the enhanced and northward-extending Bermuda High and the sharp, southward-reaching trough in the eastern United States.[8]

As a tropical storm or hurricane moves northward steered by a trough, it is also moving into an environment of higher wind shear and colder ocean water, both of which lead to hurricane weakening and dissipation. Additionally, the air on the west and/or north side of the trough tends to be quite dry

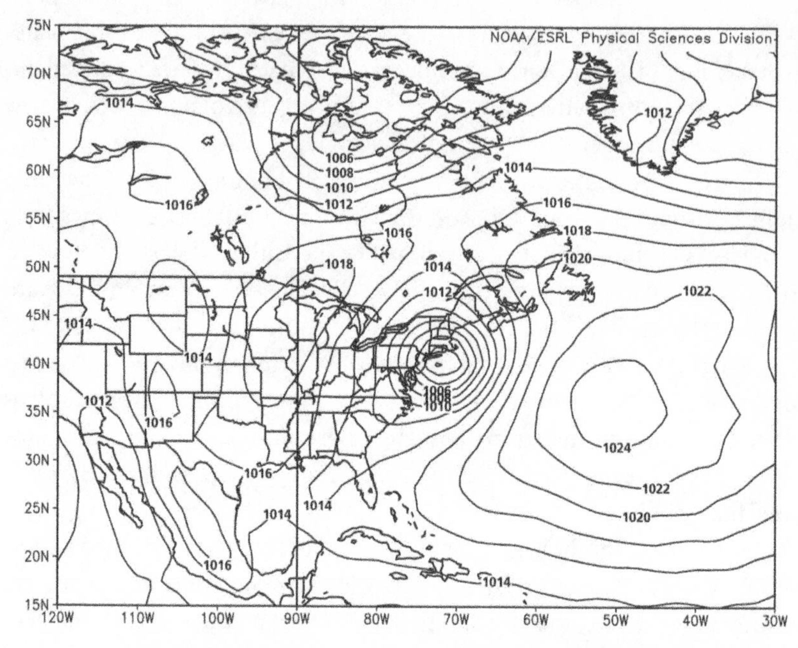

Composite surface pressure pattern (in hectopascals) for a typical New England hurricane strike constructed by combining the sea level pressure conditions during six post-1950 hurricanes with similar New England tracks regardless of their intensity: Carol (1953), Carol (1954), Edna (1954), Donna (1960), Gerda (1969), Gloria (1985), and Bob (1991). The pattern shows a hurricane centered just south of Long Island and a large Bermuda High dominating the Atlantic Ocean. (The composites were built using NOAA's Earth System Research Laboratory's 6-Hourly NCEP/NCAR Reanalysis Data Composite website.)

(and cold) and its intrusion into the hurricane further leads to a disruption of the storm's moisture structure and hence results in further weakening. For a storm to maintain its strength well into the middle latitudes—for it not to weaken as fast as it should in such a normally unfavorable environment—the hurricane must align with the trough in such a way that it can use a new source of energy not available in the tropics. The energy is provided by the temperature contrast on the boundaries of the trough, in what Pierce called the "energy of air mass distribution."[9] It might seem that any position along the trough would provide such extra energy, as the temperature contrast occurs throughout, but the hurricane must simultaneously be positioned under an enhanced area of upper-air divergence for further strengthening to occur. This generally occurs to the east of a trough axis, but there is a jet stream feature that can further enhance the divergence. A close inspection of the composite jet stream occurring during New England hurricane landfalls shows a maximum in speed, known by meteorologists as a jet streak, over the

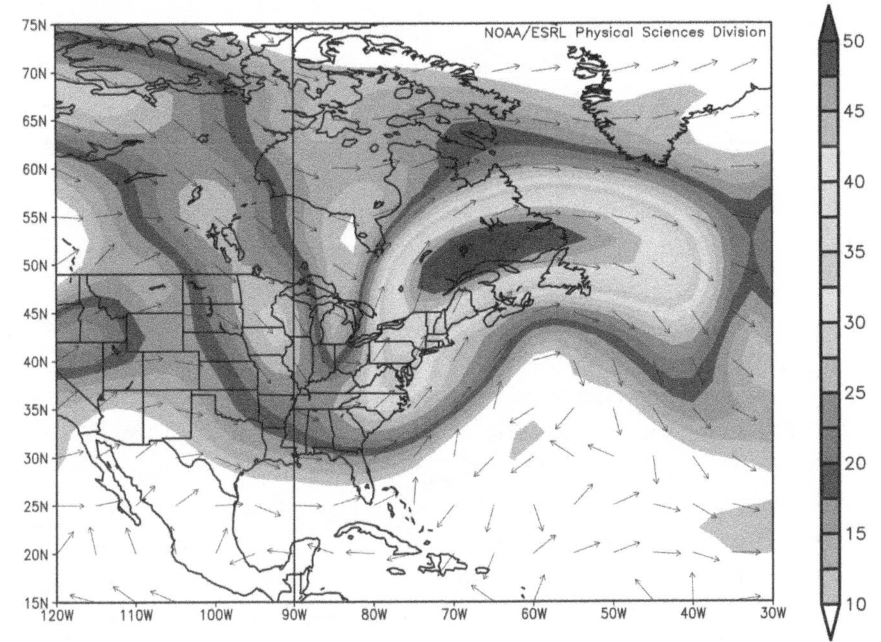

Composite upper wind pattern involved in steering a typical New England hurricane constructed by combining winds near the tropopause, at the 250-hPa pressure level, for the same six hurricanes and also using NCEP/NCAR Reanalysis Data. The vectors show wind direction, and the shading shows wind speeds in meters per second. The pattern shows a deeply troughed jet stream that would steer storms northward or northeastward toward the region.

eastern half of the trough. Parts of a jet streak, the so called "right-entrance" and "left-exit" regions (in this case the bottom left and top right quadrants of the wind maximum, respectively), provide extra upper divergence to further enhance (deepen) any favorably positioned surface low.[10] The right-entrance region of the jet streak appearing over the New England hurricanes composite is especially well positioned for strengthening a northward-moving hurricane. There is no way to know exactly from the available 1938 data, but the timing and motion of the Hurricane most likely put it in a good position to tap the enhanced energy source provided by the right- entrance region of a jet streak on the eastern side of the existing trough. This switch in energetics also marks the transition of the storm from a purely tropical cyclone to an extratropical system, as will be described in more detail in the next chapter.

"The Long Island Express"

Not surprisingly, the currents that move hurricanes around the globe determine not just their direction but also how fast they move. The 1938 Hurricane was extraordinary in this regard, moving faster than had ever been observed or even thought possible at the time. Cape Verde hurricanes like this one first move at a leisurely 10 to 15 mph or sometimes even slower while crossing the Atlantic within the weak-steering background flow found in the deep tropics. As recurvature into higher latitudes occurs, it is not unusual for them to then accelerate to speeds in the 20- to 30-mph range or even faster, as the background steering flow is generally stronger. This is the case even if the storms are not interacting directly with the jet stream.

The enhanced ridge over the North Atlantic combined with the deep trough over the eastern United States in September 1938 produced a large horizontal pressure difference between the continent and the ocean. The correspondingly large pressure gradient force had thus set in motion a very strong ambient flow. Once the storm approached and coupled with the eastern side of the trough, it moved into extremely strong steering currents. In his 1939 article on the meteorology of the storm, Pierce shows this clearly in his analyses. North-south-oriented isobars very close to each other show a strong northward-steering flow that would have produced a fast-moving hurricane. A careful look at the pressure pattern at 10,000 feet during the morning of landfall shows a difference in pressure of 0.5 inch of mercury (approximately 16 hPa) between central Pennsylvania and Cape Cod. The pressure difference across the same region for both the previous and following days is only 0.1 inch (approximately 4 hPa). It is clear, then, that the pressure gradient intensified immensely just as the Hurricane was approaching land—as the trough deepened both in intensity and southward extent.[11] The powerful background flow easily took the storm from the vicinity of Cape Hatteras in the morning to its Long Island and southern New England landfall in just a few hours.

How fast was the Hurricane actually moving? Even though the original summary of the storm appearing in the *Monthly Weather Review* (MWR) reported that its forward speed was 50 mph as it approached New England on September 21, speeds of up to 70 mph and a distance traveled of 600 miles from morning to afternoon have been attributed to the Great Hurricane since then.[12] However, a careful examination of the estimated position at six-hour intervals in HURDAT/2 reveals that its forward speed, though still extreme for a hurricane, was not as high as popularly thought. One can calculate the distance between two latitude–longitude pairs (while keeping in mind the curvature of Earth) and divide by six to obtain an average speed in miles per hour.[13] Applying this analysis to the storm positions in both

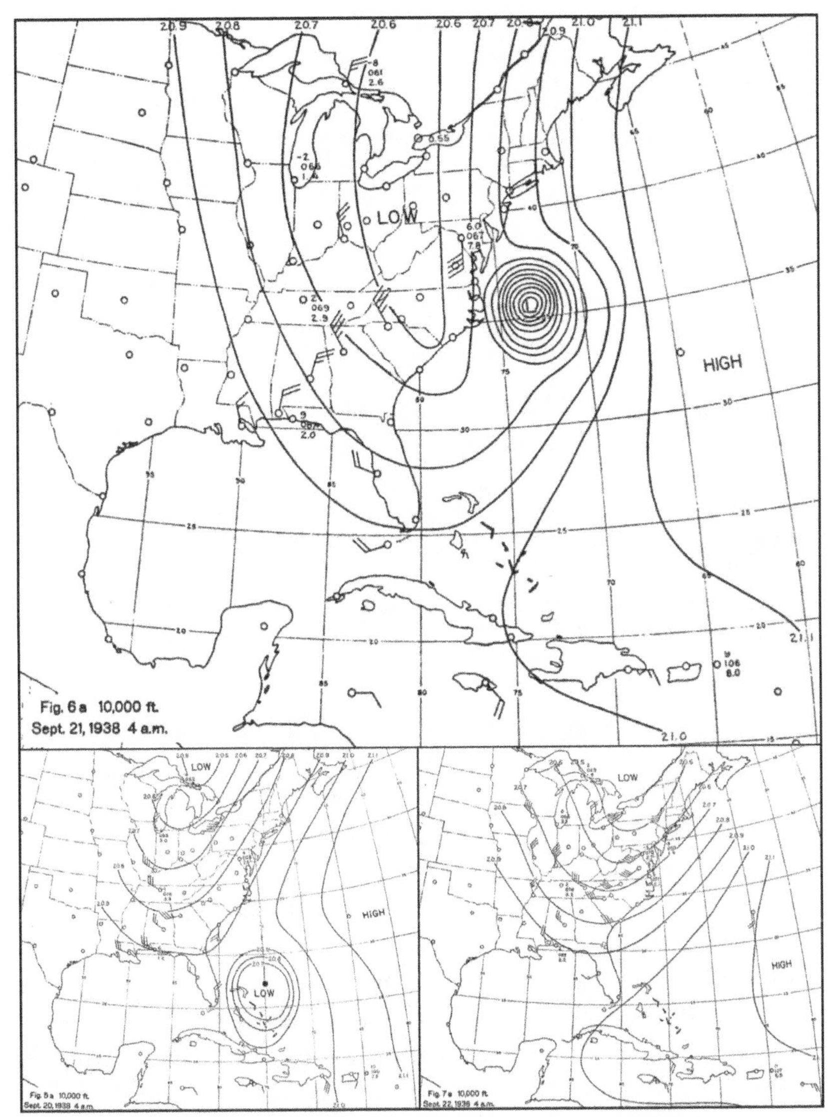

The top panel shows atmospheric pressure conditions at 10,000 feet during the morning of the Hurricane. The bottom panels show the equivalent map for the day before and the day after the Hurricane. The lines are isobars in inches of mercury (every 0.1 inch) as was the meteorological practice at the time. The strength of the flow is proportional to the pressure changes (isobars closer together mean stronger winds). On September 21, 1938, the pressure trough spanning the eastern United States deepened in north–south extent and strength, and the Hurricane was steered straight northward by the resulting very strong steering flow. All three maps are hand-drawn analyses based on the scarce upper-air observations available at the time, as presented by Pierce's 1939 detailed study a year after the storm.

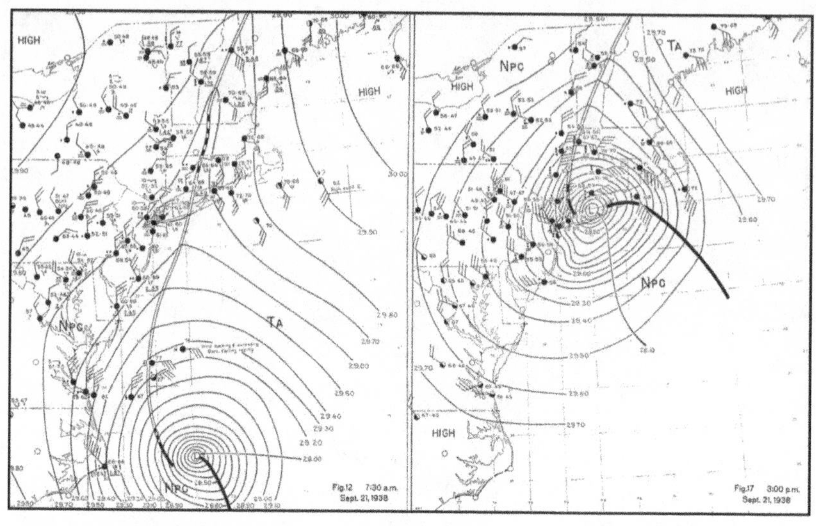

Surface observations and pressure analyses show that the center of the Hurricane was in the vicinity of Cape Hatteras at 7:30 A.M. on September 21 and over Long Island and southern New England by 3 P.M. The lines are isobars in inches of mercury, drawn every 0.1 inch. These are also from Pierce (1939).

the original and revised datasets, the forward speeds obtained are closer to a maximum of 50 mph (as shown in Table 6.1). The six-hourly positions represent a best estimate based on the best available information (especially that resulting from the storm's reanalysis); therefore, it is perfectly possible that the storm did travel somewhat faster, say 52, 55, or maybe even 60 mph, for short periods of time, but 70 mph would most likely be an overestimate.

Fifty miles per hour is, nonetheless, still extremely fast for a hurricane at these latitudes. An examination of storms recurving in the western Atlantic since 1980 shows that the average forward speed between the latitudes of Cape Hatteras and southern New England is 20 mph, and close to 30 mph as approaching 40°N.[14] Only one storm moved toward the southern New England latitudes during that period at speeds above 40 mph, Gustav in 2002.

The consequences of the extraordinarily fast forward motion were many. First, the wind speeds accompanying the Hurricane, the ones felt during landfall, were much stronger than the storm would have been able to produce on its own; they were a combination of both its rotation and its forward motion. This is always the case, but the effects are much more dramatic with a fast-moving hurricane. As will be described below in more detail, to the right of the storm's direction of motion, rotation and forward motion combine to produce much stronger winds. Additionally, there was less time for

TABLE 6.1. Forward Speed of the Hurricane

Date	Time[a]	Original HURDAT Eye Position	Translation Speed[b,c]	Revised HURDAT2[d] Eye Position	Translation Speed[b,c]
Sept 20	0Z (7 P.M. Sep 19)	25.0°N, 72.7°W	15.5	25.2°N, 72.4°W	14.3
	6Z (1 A.M.)	25.9°N, 73.6°W	14.0	25.8°N, 73.4°W	12.3
	12Z (7 A.M.)	26.7°N, 74.3°W	11.7	26.7°N, 74.3°W	13.9
	18Z (1 P.M.)	28.0°N, 74.8°W	15.8	28.0°N, 74.8°W	15.8
Sept 21	0Z (7 P.M. Sep 20)	29.8°N, 74.9°W	20.8	29.8°N, 74.9°W	20.8
	6Z (1 A.M.)	32.2°N, 74.4°W	28.1	32.2°N, 74.4°W	28.1
	12Z (7 A.M.)	35.2°N, 73.1°W	36.7	35.2°N, 73.1°W	36.7
	18Z (1 A.M.)	39.0°N, 73.0°W	43.8	39.3°N, 72.9°W	47.2
Sept 22	0Z (7 P.M. Sep 21)	43.3°N, 73.1°W	50.7	43.4°N, 73.1°W	47.2
	6Z (1 A.M.)	45.3°N, 73.5°W	22.1	46.5°N, 74.5°W	37.5
	12Z (7 A.M.)	47.3°N, 77.0°W	36.1	47.7°N, 77.3°W	25.9
	18Z (1 P.M.)	45.4°N, 79.1°W	27.5	47.0°N, 77.8°W	9.0

[a] Universal time coordinates expressed in Z or Zulu time are commonly used in meteorology. (EST is five hours behind Z time.)

[b] Speed is in mph.

[c] The speed was calculated by using the distance between the latitude–longitude pairs using a great circle calculation divided by six hours.

[d] HURDAT2 contains position revisions resulting from the reanalysis of the 1938 hurricane season.

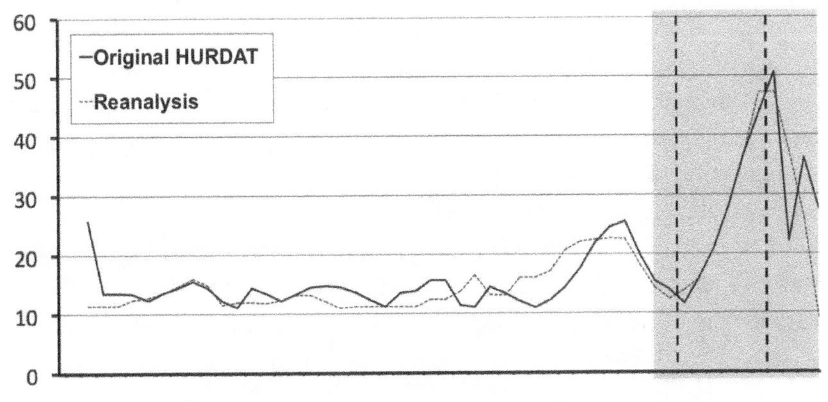

Forward speed (in miles per hour) of the 1938 Hurricane throughout its lifetime. The shaded portion of the graph corresponds to the data shown on the accompanying table, from the evening of the 19th, to midday of the 22nd as the storm dissipated. The dashed lines mark the times of recurvature and landfall. As the storm turned, it continued accelerating until it hit land on the 21st.

the unfavorable environment to take its toll on the transitioning hurricane, meaning that the storm probably avoided weakening too much while at the same time positioning itself more favorably with respect to the trough to tap into the new energy source. Lastly, this very fast-forward speed is potentially one of the biggest factors that contributed to the lack of warning from the Weather Bureau. It certainly affords the forecasters the benefit of the doubt. One might assume that if the Hurricane had not been moving as fast as it was, perhaps Mitchell and his team would have had time to revert their oversight and issue appropriate hurricane warnings for the Northeast. But of course, this was not the case. The Great Hurricane raced northward, covering roughly 400 miles from early morning until the afternoon, and in the process earned the nickname of "The Long Island Express."[15]

Daylight Saving Time and the Storm

It is easy to find contradicting timing information for landmarks such as the time of landfall or the time that a specific location was affected. Given that the time difference in the conflicting reports is always one hour, it is intuitive to suspect that the time listed in some reports is in Eastern Standard Time (EST) while it is in Eastern Daylight Time (EDT) in others. Further research, however, reveals that there are a number of intricacies to the story, and the Great New England Hurricane provides an opportunity to briefly explore the history of Daylight Saving Time (DST), which simply consists of advancing the time (usually by one hour) during the summer and adjacent months to take advantage of extra daylight during the early evening. All locations relevant to the story of the Hurricane are in the Eastern time zone; therefore, the use of DST means that the local time would have been reported in EDT.

It turns out that in 1938 the nation was living through what can only be called a period of time anarchy. Many states opted for using yearlong standard time, but some states or individual cities (or even businesses) were adamant about the benefits of moving their clock forward to gain an extra hour of daylight in the evening during the summer months. The specific start and end time for DST usage was also determined independently by each location, and not surprisingly, time-keeping became ridiculously complicated, especially for those traveling between and through locations with different time conventions.

The determination and standardization of time has a long, complicated, and fascinating history of which only a few highlights are given here.[16] The main common characteristic during the ancient times and the Middle Ages was the prominent use of local time, which was based in one way or another on the hours of daylight and the local position of the sun in the sky. Over time, as travel became more common with the advent of trains, the need for some kind of standardization became apparent. In 1883, railroads commenced the use of standard time by defining time zones for the areas within their operation. Consequently, it was common for people to follow both the railroad time and their natural local time. Very soon, it became clear that there were benefits to standardizing beyond coordinating transportation and thus a large part of the world moved toward time standardization. In 1884 the International Prime Meridian Conference in Washington, D.C., adopted the system of time zones still used today around the world. Interestingly, the United States would not jump on board with the new international agreement until a few decades later.

DST has a convoluted and quite contentious history in its own right.[17] Its use has been heralded and opposed by many individuals, special interests, and regions. The earliest seed for the concepts came from none other than Benjamin Franklin, who at 78 years old served as American Minister to France. He wrote a witty letter essay to the *Journal of Paris* in 1784 where, half humorously and half seriously, he described how daylight was being wasted, focusing on how sunlight was so much cheaper than candlelight. Franklin's idea, however, did not involve moving the clock forward but instead called for making as much noise as possible (ringing church bells and, if that did not work, firing cannons in every street) to "wake the sluggards effectively" and force them to take advantage of natural light.[18] The true champion of the modern concept of DST, most commonly referred to then as "summer time," was British home designer and builder William Willett, who unfortunately did not live to see his revolutionary idea take effect.[19]

It was not until 1918 that the United States adopted the use of DST after various European nations had done so during World War I. At this time it also finally adopted the Prime Meridian agreement of 1884, defining standard time zones for the country. Americans, however, were divided in their support of moving the clock forward during the summer. The practice was especially unpopular with farmers, who claimed that their tasks were mostly driven by the sun rather than the clock and that now they would be out of synch with railroads and markets, hurting their business. The law mandating nationwide use of DST was repealed after the war (in 1919) by Congress, which in the end won the political ping-pong versus President Woodrow Wilson (the repealing law was passed twice, the president vetoed it twice, and the overruling of the veto by a two-third majority finally passed the second time around). During World War II, a new law mandating DST was once again put in place.

TABLE 6.2. Use of Daylight Saving Time in 1938

Florida	EST
Washington, D.C.	EST
Long Island, NY	EDT
Rhode Island	EDT
Connecticut	EDT
Massachusetts	EDT
New Hampshire	EDT
Vermont	EST

NOTE: Local time used on September 21, 1938

In between the two World Wars, there was no federal regulation that either mandated or prohibited the use of DST. The standard time zones remained in place, but states, cities, businesses, or institutions individually decided if and exactly when to observe the summertime forward shift of one hour. This debate was still going on in September 1938. Table 6.2 shows the usage of DST throughout locations relevant to the story of the Hurricane. To make matters somewhat more complicated, most hurricane reports and other related documents at that time list time without specifying if it is EST or EDT. The Weather Bureau advisories and reports were done in standard time, which was officially used for federal government business as well as throughout Washington, D.C., where the Bureau headquarters was located. Most of the locations relevant to the passage of the Hurricane used EDT, as the Northeast was consistently the most enthusiastic region for the practice, although not all of them jumped in right away. For example, in northern New England, "Live Free or Die" New Hampshire at first refused to go along with the majority of its neighboring states. In southern New England, it was Connecticut that resisted for a while, but by the late 1930s both states had joined in the trend. Most of the local storm accounts and reports, then, use EDT.

With the Weather Bureau using EST and most of the Northeast using EDT, confusion among the general population might have ensued when storm warnings came through (perhaps a moot point in the case of the Great New England Hurricane, since no formal hurricane warnings were ever issued). This would not have been a problem with the Jacksonville Office, whose most interested audience was also in Florida and using the same (standard) time.

Even though Franklin's original idea was inspired by financial gain, most of the arguments highlighting the benefits of using DST emphasized the

TABLE 6.3. Hurricane Landfall Statistics

	Long Island	Connecticut
Location	40.7°N, 72.9°W (near Brookhaven)	41.3°N, 72.9°W (near New Haven)
Time	2:45 P.M. EST	3:40 P.M. EST
Pressure	941 hPa	946 hPa
Wind speed	105 kt (120 mph)	100 kt (115 mph)
Radius of maximum winds[a]	40 nmi (45 miles)	40 nmi (45 miles)
Outer pressure[b]	1011 hPa	1011 hPa
Storm width[c]	700 mni (800 miles)	700 nmi (800 miles)
Forward speed	41 kt (47 mph)	41 kt (47 mph)

NOTE: Data originally from preliminary reanalysis report (Landsea 2008), now in HURDAT2.

[a] RMW—Radius of Maximum Winds—distance of strongest winds from center of storm.

[b] OCI—Outer Closed Isobar—pressure in the outskirts of the storm.

[c] Two times ROCI—Radius of Outer Closed Isobar—estimated width of outer part of storm.

extra time for recreation in the evenings, health improvements from additional outdoor time, and general quality-of-life enhancements. What's more, during the wars, the reduced energy usage idea fit very well with the austere needs and sentiments of the government and the population, who felt that the savings obtained from an hour less of electrical lighting were worth whatever inconveniences the change might cause. The 1938 Hurricane provided an opportunity to take advantage of the extra daylight in a way that was never anticipated. DST allowed the affected communities an extra hour of daylight at a more convenient time for rescue, relief, and repairs. Most locations in the area were scheduled to end the use of DST on September 25—the last Sunday of September and only four days after the arrival of the storm. However, the governors of the affected states decided to extend it for an extra week as an emergency measure.[20]

For now, regardless of time-keeping practices, the Hurricane was still rapidly approaching. It was a stormy night in Hatteras, North Carolina, which experienced strong winds starting shortly after midnight on the 21st and continuing through the early morning (even though the center of the Hurricane stayed offshore). The storm then continued its unforeseen advance straight for land, and conditions deteriorated quickly after a relatively calm morning. Wind and rain started to pick up in Long Island and southern New England early in the afternoon. The center of the storm reached the southern coast of Long Island at 2:45 P.M. (EST), then crossed the Long Island Sound, and barely slowed down before a second landfall in Connecticut less than an hour later at 3:40 P.M. (Table 6.3 contains more landfall statistics.) The 1938 Hurricane thus arrived in unsuspecting and unprepared New England with its many accompanying hazards.

The Storm's Hazards

Jurakan, the old storm deity of the Taíno natives of the Caribbean islands, was feared for good reason; hurricanes bring along a frightening assortment of dangerous hazards including heavy rain, strong winds, high waves, coastal and inland flooding, and sometimes even tornadoes. Individually and in combination with each other and with the local environment, these elements are capable of generating immense devastation (see Table 6.4). Many of these hazards played a prominent role in the destruction caused by the 1938 Hurricane.

No Tornadoes?

Tornadoes, meteorologically defined as columns of violently rotating air extending from cloud to ground, are much smaller entities than tropical cyclones—although the intensity of their winds can surpass that of an extreme hurricane. A large violent tornado can be one or two miles wide with accompanying surface winds well above 200 mph, and typically lasts on the order of tens of minutes up to a couple of hours. In comparison, a large violent hurricane is 1,000 miles wide or even larger and can last for several days, but its sustained winds do not reach 200 mph. Of course, the mechanisms for their formation and evolution are also completely different. The most violent tornadoes originate from strong, rotating thunderstorm clouds known as supercells that sometimes form within the most favorable areas of extratropical cyclones, although other atmospheric situations conducive to vertically shifting wind directions can also produce them.[21] It is very common for the landfall of hurricanes to be accompanied by tornadoes of the less violent variety, mostly within the rainband thunderstorms preceding the center of the storm and within the eyewall. These tornadoes can produce localized areas of additional damage superimposed on the damage created by the hurricane as a whole. Most hurricanes making landfall in the United States are indeed accompanied by at least one tornado, and sometimes many more.[22] Hurricane Ivan, for example, produced a total of 127 (the largest number ever observed to date) during its passage through the Gulf and East Coasts in 2004.[23]

TABLE 6.4. Hazards Accompanying the 1938 Hurricane

Hazard	Factors Involved
Excessive rain	-
Damaging wind	-
Downed trees	rain + wind
Landslides	rain + mountains
Inland flooding	rain + rivers
Storm surge	wind + coast
Storm tide	storm surge + lunar tide
High-breaking waves	Storm tide + wind + coast

There is no report or indication of any tornadoes accompanying the 1938 Hurricane. However, it was only during the 20 years before the storm that tornadoes inside hurricanes were finally detected, and only twice had the phenomenon been documented then, in 1919 and 1924.[24] It is reasonable to suspect that the 1938 Hurricane might have spawned a tornado or two that went undetected, but so far there is no direct or indirect indication of such occurrence within any scientific or popular account of the storm.

Hurricane Rain

Excessive rainfall is, of course, a common tropical cyclone hazard. Even if winds are not strong enough for a system to be designated a hurricane, a storm can still bring tremendous amounts of rain in a short period of time to locations within its path. The most intense rain tends to occur within the eyewall, but significant amounts can also be produced in the rainband thunderstorms. Anywhere from 2 to 20 or more inches are possible in a time period of a few hours to a couple of days, depending on the size of the system and how quickly it is moving. A slow-moving storm is more likely to bring more precipitation to a location, as it takes several hours for it to pass. This was not the case with the 1938 Hurricane, which was moving very fast at time of landfall. In fact, all of the rain from the Hurricane fell within half a day on September 21. A standard Hurricane rainfall analysis by the Weather Prediction Center (formerly named the Hydrometeorological Prediction Center) averages the available daily rain totals from September 19 to 22 to depict the total storm rainfall.[25] This summary shows amounts from 1 to 10 inches along the East Coast and across New England. The approach gives us a general idea of the rainfall throughout the region, but it does not purely show the Hurricane rainfall. It also partially captures a preceding four-day

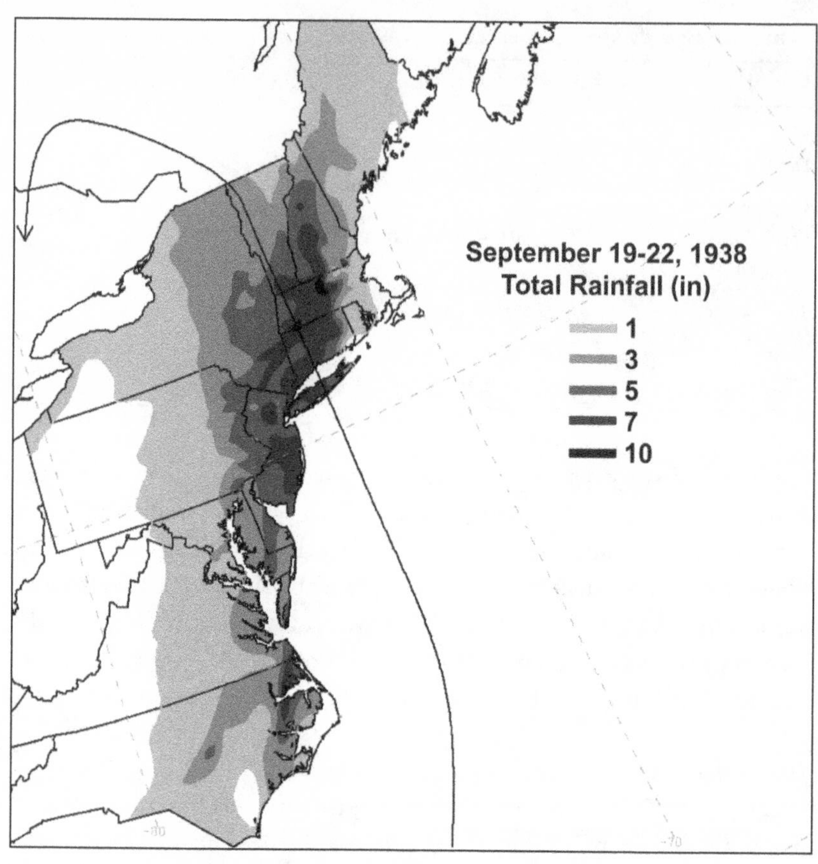

Tropical Cyclone Rainfall Data for the 1938 Hurricane adapted from the Weather Prediction Center. The map combines rainfall observations (available from the National Climatic Data Center) for September 19–22. All rainfall directly related to the Hurricane came during the 21st (and maybe a little more on the 22nd for the northernmost locations). The very significant preceding rainfall lasted approximately from the 17th to the 20th. This means that the totals shown here are a combination of a portion of the preceding event and the Hurricane.

rain event. The region had been soaked with tropical moisture riding up what most likely was a stationary front hovering around the area in association with the previously mentioned deepening trough. The storm was thus preceded by days of heavy rain, which might have been what is, since 2010, known as a Predecessor Rain Event (PRE), causing many rivers and streams to be already at or near flood stage (see Chapter 7 for more about PRE and the effects of the excessive rain).

Detailed 24-hour rainfall records are available from the National Climatic Data Center (NCDC), but discerning exactly what portion of the rain came directly from the Hurricane can get a little trickier than one might think. All of the Hurricane rain came during the afternoon and evening of September 21, and maybe a little more during the early hours of September 22 for some of the most northern locations. The majority of the rainfall records are identified as afternoon measurements, which would have been at the height of the violent weather. If observers ventured out to check their rain gauges in the middle of the storm, then the Hurricane rain total would be split between the reports on the 21st and the 22nd. Unfortunately, the number reported for the 21st would also include rainfall for the previous evening and even afternoon, which means it might include some of the preceding rain event. A significant portion of the stations' reports came from morning measurements of the previous 24 hours, which suggests that the Hurricane rainfall would be recorded on the 22nd rather than on the 21st, but it is unclear from the accompanying documentation if the numbers were moved to the previous day to account for this. However, for most stations one can discern which value is meant for one day or the other by carefully looking at the recorded amounts. Finally, a smaller number of reports came from Weather Bureau stations, where the measurements were done from midnight to midnight. In this case, most of the Hurricane rain would appear on the record for the 21st. By adjusting for the different timing nuances of the observations and keeping in mind some of the uncertainties, the Hurricane rain amount can be extracted from the historical records in a more accurate way than simply choosing the range of days bookending the storm and adding the numbers.

In general, from the early afternoon to the late evening, most locations directly affected by the Hurricane received anywhere from one to seven inches of rain as the storm made its way through. Table 6.5 shows only a few sample locations per state, the highest numbers and the biggest cities. Overall, locations in Connecticut experienced the highest rain totals, although the single highest number was measured in Hillsboro, New Hampshire (with 7.36 inches, although it is possible that some of that rain came during the previous evening), and different degrees of high amounts occurred in every New England state. One glaring omission from the rainfall table and this discussion is Gardner, Massachusetts, which has formerly been identified in modern accounts of the storm as the location that received the most rainfall from the 1938 Hurricane, with a total of 12.77 inches. When taking into account the preceding rain and the timing of observations, it turns out that most of this rain in Gardner fell before the Hurricane and cannot be attributed directly to it. The rainfall in Gardner during the 21st was a much less impressive 1.63 inches. (The NCDC records report 11.14 on the 20th, but

TABLE 6.5. Rainfall Exclusively Due to the Hurricane

Connecticut	
Burlington	6.53"
Storrs	6.52"
Lake Konumoc	6.50"
Norwalk	6.42"
Bridgeport	5.08"
Hartford	3.22"
New Haven	1.72"

Vermont	
Wilmington	6.90"
Whitingham	6.68"
Readsboro	6.15"
May's Mill	5.72"
Searsburg Station	5.71"
Burlington	2.85"

Maine	
Middle Dam	4.22"
Brassna Dam	2.95"
Greenville	2.69"
Portland	0.00"

Rhode Island	
Block Island	1.00"
Kingston	0.19"
Providence	0.17"

Massachusetts	
Hoosac Tunnel	7.01"
Springfield	5.81"
Shelburne Falls	5.79"
Westfield	4.73"
Heath	4.70"
Stockbridge	4.42"
Boston	0.10"

New Hampshire	
Hillsboro	7.36"
Franklin	5.87"
Lakeport	5.36"
Plymouth	4.52"
Bethlehem	4.29"
Pinkham Notch	4.26"
Concord	2.79"

SOURCE: From September, 1938 climate records for the New England region from NCDC.

as a two-day total—for the 19th plus the 20th, all too early to be hurricane rainfall.) There is much less uncertainty involved in calculating the total rainfall from both the preceding rain event and the Hurricane, for which Gardner has one of the highest, amounts, with a combined total of 14.52 inches. Rather than the Hurricane alone, this combination of the two events, which will be addressed in more detail in the next chapter, is what truly caused the hazardous effects directly associated with the excessive rainfall.

As expected, large amounts of rain can cause flooding of inland bodies of water, which was one of the great problems during the passage of the 1938 Hurricane. Excessive rain can also cause landslides in mountainous areas and soften the terrain enough for the wind to knock down trees more easily. The latter was also a huge problem in 1938. Tens of millions of trees were blown down throughout much of New England, leaving behind an unrecognizable landscape and prompting the mobilization of *the* largest timber salvage operation in history (see Chapters 7 and 8 for more about the tree damage and cleanup).

Hurricane Wind

The winds accompanying a hurricane do not blow uniformly throughout the storm. As previously described, the lightest winds are inside the eye—which is also the area of lowest atmospheric pressure. The strongest winds, set in motion by the immense pressure difference (between the eye and its immediate surroundings), are found within the eyewall. From there, they decrease sharply at first and then gradually toward the outskirts of the storm until they blend with the surrounding background flow. Short-duration gusts of wind can also be experienced throughout the storm, especially within the spiral rainbands extending outward from the center and, not surprisingly, also within the eyewall. These individual gusts can be much stronger than the sustained winds (see Chapter 1) normally used to describe the strength of a storm.

As briefly described above, the winds experienced during the passage of a hurricane are also determined to a large extent by how quickly it is moving along its track. On the right side of the storm (with respect to its motion), this forward speed is in the same direction as the rotation (which is counterclockwise in the Northern Hemisphere), and the winds experienced at the surface are a simple sum of the two. To the left, forward speed is opposed by the rotation, and wind speeds are lower than they would be if the storm was not moving at all. Since the Hurricane was moving so fast as it approached land (between 40 and 50 mph), this created an especially strong effect: there was a high degree of asymmetry between the east and the west sides of the northward-moving storm, with the winds on the east side of the eyewall being significantly stronger. In general, around the time of landfall, the winds experienced over eastern Long Island, eastern Connecticut, and Rhode Island were roughly in the 60- to 100-mph range. On the other side, over western Connecticut, western Long Island, and New York City, they were more in the 20- to 50-mph range.[26] The winds just specified have been adjusted to one-minute averages to match the current definition of sustained winds; however, in 1938 it was common to report five-minute-averaged winds.

Some of the strongest winds experienced during the Hurricane occurred at higher elevations rather than at coastal locations. For example, Mount

Washington, in the White Mountains of New Hampshire and the highest peak in the northeast United States (at an elevation of 6,288 feet), reported average winds of 118 mph and gusts of up to 163 mph. The strongest wind report came from the Blue Hill Observatory, just south of Boston, at the highest elevation in the immediate coastal region (635 feet). Their most accurate anemometer was destroyed by the strong winds shortly after 5 P.M., missing the strongest winds of the evening, but a more sluggish three-cup anemometer was at hand and measured a five-minute wind of 121 mph and gusts of up to 189 mph. The Blue Hill Observatory, boasting the oldest continuous record of weather observations in North America, produces perhaps the most detailed observer descriptions of historical data that can be found, and the observations for the 1938 Hurricane are no exception.[27] Many of the descriptions for September 21 note the wind and its effects on the surrounding area:

> Wind increased rapidly after 2:00 p with approach of hurricane. At 3:21p, blades on windmill . . . blew off after recording 37.2 m/sec. 3:14p, lights failed; 3:30p, apple tree in yard split; 3:30p, spray being lifted off surface of Ponkapoag (a pond 1½ miles away, and a little under ½ mi across) possibly 30 ft in the air. 3:40p, covers of coal bin carried off and split . . . 5:20p, odor of bruised oak and cherry leaves very noticeable outside. About half of trees on south side loosened and partially uprooted. Many totally blown out of ground, or snapped across trunk. Leaves frayed and whipped off all trees, most of which have bark worn off of minor branches; and large branches snapped off.

> The wind on Blue Hill reached 60 mi/hr shortly after 3 P.M. and remained continuously above that velocity from 3:35 to 6:45 P.M. From 4 to 5 the wind averaged 83 miles per hour. Fifty miles of wind went by in only a little more than half an hour, at 94 mi/hr. In three five-minute periods, at 5:05, 5:20 and 6:12, the wind reached 111 miles per hour . . . Though the wind kept up a high velocity it was punctuated by great gusts, which would last for a few seconds to a minute. The most sensitive recorder at Blue Hill a French windmill anemometer started to disintegrate when registering a 5-minute velocity of 80 miles, the gust that broke it being well above 100; so we have to use the cruder record from the heavier 3-cup anemometer for such details as to the force of the gusts . . . From this record it appears that there was a passage of 7 miles of wind in 2½ minutes at 5:59 P.M. and one of 4 miles in a little over a minute at 6:15. With due allowances for instrumental errors it appears that these gusts were approximately 173 and 186 miles per hour.

TABLE 6.6. Five-Minute Wind Reports from Original MWR Report

Albany, NY	42 W
Block Island, RI	82 SE
Boston (airport), MA	73 S
Burlington, VT	47 S
Concord, NH	56 SE
Eastport, ME	32 SE
Hartford, CO	46 NE
Nantucket, MA	52 SE
New Haven, CO	38 NE
New York City	70 NW
Northfield, VT	47 S
Portland, ME	43 S
Providence, RI	87 SW

SOURCE: September 1938 MWR storm report by I. R. Tannehill.

NOTE: Winds in miles per hour—five-minute average—nonstandard anemometer heights.

Stating specific observed wind speeds can become a little tricky for more than one reason. The current standard for sustained wind reporting in the United States is a one-minute average.[28] The wind reports at the Weather Bureau and other observing stations in 1938 specify a "5-minute interval" or "5-minute velocity" (see Table 6.6), which would translate to a somewhat higher one-minute average (roughly about 6% higher).[29] For example, the five-minute, 121-mph Blue Hill observation would correspond approximately to a one-minute, 128-mph sustained wind in modern convention. Moreover, the observations sometimes reported as gusts were one-minute averages, which would make them sustained winds today. There is also the matter of the altitude at which observations were taken—in our case specifically the height of the anemometer. The strongest winds in a hurricane are near the surface, but not right at the surface, since they are slowed down by friction with the ground. The maximum wind is therefore toward the top of this layer of atmosphere affected by friction, usually a half- to one-kilometer above the ground. This means that surface measurements by weather stations do not capture these highest hurricane wind speeds, which is not a problem because the most relevant speeds are those experienced on the ground by people and structures. The current standard height for surface wind speed measurements is 10 meters (approximately 33 feet), but in 1938 there was no such standard. A brief survey of detailed instrumentation information for weather stations in the Washington, D.C., area, for example, shows that anemometer heights ranged wildly at the time.[30] To properly compare and analyze the winds, one must know the anemometer height of all the wind observations and convert them by an appropriate factor to the standard 10-meter height.[31] Regardless of these nuances, however, the highest reported wind speeds on the coast appear to have come from Providence with 76 knots (87 mph). Unfortunately, there are no wind reports from anywhere in Long Island to the right of the path of the center of the storm at the time of landfall. Marine stations in the vicinity, however, reported even higher winds. Fishers Island, New York, just south of the Connecticut–Rhode Island border, reported 95 knots (110 mph) just before the anemometer failed. A similar story came from the Coast Guard station in Westerly, Rhode Island, where the observer estimated a maximum of 105 knots (120 mph). The official estimated landfall wind speeds resulting from the hurricane reanalysis project (see Chapters 1 and 2), are 105 knots (120 mph) for the first landfall in Long Island and 100 knots (115 mph) for the second landfall in Connecticut.[32]

Storm Surge

The winds of a hurricane and the ocean water working together make a powerful combination that can result in a variety of hazardous effects. Persistent strong winds blowing over a large expanse of water produce large waves that are a hazard both in the open ocean (to any ships that happen to be in the vicinity) and on the shore (where they batter any structures that happen to be within their reach). There are also less obvious wind-driven hazards. As the wind pushes the surface water, underwater currents are set in motion. In shallow waters, these secondary currents push the sand on the shallow bottom, creating underwater sandbars that can be large enough to block water from going back toward the ocean until its weight causes the bar to breach. The water then returns in a restricted, strong flow out to sea. These sometimes invisible and unexpected currents are known as rip tides or rip currents, and they can be very dangerous to swimmers, whose first instinct is to swim against the current toward shore instead of the much safer cross-current swimming parallel to the shore.

Without doubt, the most dangerous consequence of the wind–ocean interaction is the storm surge, simply defined as the abnormal rise in ocean water height accompanying a hurricane or other intense storm.[33] The strong and persistent blowing of the hurricane winds pushes the ocean surface water. While the hurricane is still over open water, the resulting sea level elevations are slight and most of the energy is spent driving relatively shallow currents below the surface. When the storm comes closer to land and interacts with the shallow depths leading to the shore, then the large increases in sea heights occur. There is also a small storm surge component due to low pressure within the eye. The water level is literally pulled upward at the center of the storm, analogous to when liquid is sucked up a straw. However, this is a much smaller effect and mostly negligible except for very intense storms. (The rise is roughly one centimeter for each hectopascal of pressure drop. An intense hurricane with a central pressure of about 900 hPa would therefore cause approximately a one-meter, roughly 3¼ feet, increase in sea level).[34] Throughout the years, the storm surge has been given various different (though not quite as accurate) names, such as storm tide and storm wave, the latter of which was most commonly used in 1938.

The storm surge experienced at a specific location for a specific hurricane depends on many factors, such as the size and forward speed of the storm, the shape of the coastline, the angle at which the storm strikes, and the depth and slope of the offshore ocean floor.[35] Needless to say predicting storm surge is a complicated matter, and the one-size-fits-all approach to each Saffir–Simpson category that was used in the past could be highly inaccurate (as briefly discussed in Chapter 1).[36] Sophisticated computer models using detailed coastal data are needed to properly forecast the expected storm surge for an approaching hurricane.[37] Even if accurate predictions are available, the storm surge is not the actual water level that will be experienced at a location. It is superimposed onto the natural astronomical tides, which can be very variable from day to day, but are also very well understood and predicted—and in general show a semidiurnal cycle with two high tides and two low tides. The actual water level observed for a storm, combining the effect of the storm and the astronomical tides, is technically termed the storm tide (not the same as in the old usage). In simple terms, the storm surge is all the extra water height that would not have been there if it wasn't for the presence of the hurricane. If making landfall toward high tide, the storm surge and the natural tide combine to produce the highest possible water levels, which can also potentially reach farther inland. On the other hand, a low tide can be seen as cancelling some of the effects of a hurricane, since the water brought in by the storm starts filling in at heights below average sea level. Thankfully, the 1938 Hurricane did not arrive exactly at high tide, although the timing was far from ideal. Most locations experienced

TABLE 6.7. Maximum Water Height above Mean Low Water

Connecticut		Rhode Island	
West Haven	16.6	Portsmouth	26.5
Stamford	15.6	Point Judith	22.1
Bridgeport	13.8	Newport	21.6
New Haven	13.8	Sakonnet	19.8
Saybrook	13.4	East Providence	19.0
Clinton	11.5	Pattaquamscott R.	18.5
		Providence	18.5
Massachusetts		Barrington	18.3
Marion	18.8	Riverside	18.3
Monument Beach	18.7		
Bourne	18.1	**New York**	
Fall River	16.1	Port Washington	18.0
Woods Hole	11.4	Montauk	16.7
Boston	11.3	East Moriches	15.7

SOURCE: U.S. Geological Survey 1940 report on the "Hurricane Floods of September 1938."
NOTE: Height in feet. Highest water height was chosen for locations with multiple reports.

the maximum storm effects ranging from about three hours (in the more southern coastal areas in southern Long Island) to about half an hour (in Boston) before the predicted high tide.[38] Nevertheless, many coastal stations for which water levels had been recorded for many years experienced the greatest water heights on record until then.

To know the exact effect of a tropical cyclone on water height, one must thus subtract the astronomical tide. For historical storms, this is not an easy task, if possible at all. The best available data usually include maximum water heights, but the exact timing is often unknown, as many are obtained, for example, from high water marks left behind after the water recedes. Most of the data for the 1938 Hurricane were obtained in this way: from high water marks at many locations along the coast. As suggested above, however, we do have some information on the timing of the maximum storm surge with respect to the high tide at a few locations.

Water height data are normally reported with respect to a specific height, referred to in the hydrology field as a datum, above which said height is measured. Various possible reference heights can be used for this purpose.[39] Today, it is very common to use Mean Sea Level (MSL) as the reference. Another common datum is height above Mean High Water (MHW), defined as the average of all high tides throughout a 19-year period (known as a tidal epoch).[40] This clearly shows how extreme the water height experienced during an event such as a hurricane can be in comparison to normal high tides for a certain location. It turns out, however, that the most complete and reliable data for the 1938 Hurricane are the height above Mean Low Water (MLW), the average of all low tides for the tidal epoch (see Table 6.7).

Historically and globally, the great majority of deaths associated with tropical cyclones are due to the sudden flooding of the ocean onto unsuspecting coastal communities in the path of the storm. In the United States, however, where advanced warnings and evacuations are commonplace, storm-surge drowning deaths are much less common these days. Inland flooding is the second overall largest killer, even though any individual inland flooding event does not normally cause as many deaths as the storm surge from an unusually catastrophic event such as Hurricane Katrina (2005).[41] In 1938, the unexpected coastal flooding brought in by the Hurricane had tragic consequences. Hundreds died along the coast—a great majority due to the sudden "storm wave," as the storm surge was then commonly termed.[42]

The Most Dangerous Side

After examining all the hurricane hazards, a pattern emerges. The heaviest rainfall is experienced in the eyewall, the strongest winds are experienced on the right side of the eyewall, and the strongest storm surge is experienced on the right side of the eyewall as well (and gradually decreases outward toward the right). It is clear that the strongest portion of a hurricane is its right side with respect to its direction of motion, and more specifically the right side of the eyewall. Mariners used to refer to this as "the dangerous right semicircle," since the ocean conditions experienced if they were caught in it were truly dangerous (the opposite side was known as the "navigable semicircle").[43]

For a northward-moving storm such as the 1938 Hurricane, this more dangerous portion is east of the eye (with the west side experiencing much less damage). Because of this, the eastern half of Long Island, the eastern half of Connecticut, most of Rhode Island, and parts of eastern Massachusetts experienced the worst of the storm's fury.

The damage, however, was not restricted to a narrow path corresponding to this most dangerous portion of the storm. The entire region experienced stormy conditions. The fact that the Hurricane was transitioning into an extratropical storm meant that, even though it would diminish in strength, it would also increase in size. The area experiencing rainfall and winds of at least gale force would expand much farther out from the center (see Chapter 7 for more about the size of the rain and wind fields due to the extratropical transition of the storm and its inland effects).

Rubble from broken buildings, splinters from broken homes, fallen trees, and debris of all kinds were left behind, in great part caused by the force of the incoming ocean water. In some coastal communities, the devastation was so complete that the areas were said to have been returned to pristine predevelopment conditions. Many dramatic stories of survival have been told throughout the decades, but many more first-hand accounts of those who lost their lives in the catastrophic storm will unfortunately never be told.

The path of destruction did not end at the coast line. The storm, now weakening—although not as rapidly as would have been expected—and becoming larger in areal extent, continued its northward path deep into interior New England.

INTERIOR NEW ENGLAND

7

Widespread river flooding and massive forest destruction

Without a doubt, the most tragic effects of the 1938 Hurricane occurred in the coastal areas, but the devastation certainly did not end there. The storm continued moving rapidly inland, and a series of towns and cities even less accustomed to hurricanes and their hazards than their southern New England and Long Island neighbors came to experience its fury.

Extratropical Transition and Its Unfortunate Timing

As the path of destruction advanced inland, the very nature of the Hurricane was now changing; it was undergoing what meteorologists call extratropical transition. At first, this might seem like a desirable turn of events, because extratropical cyclones are less intense and, you could say, less organized than tropical cyclones, but the timing of the transition in this case could not have been worse. The storm made landfall as a major hurricane, with its intense winds and strong storm surge near the center, at the same time it was starting to transition, thus expanding its wind field as it moved inland. Although weaker, it was still strong enough to cause significant tree damage, and a much broader area of already sodden land was now exposed to gale-force winds and additional rainfall.

Extratropical transition is not a rare phenomenon. Roughly half of all tropical cyclones contained in HURDAT/2 have been classified as extratropical toward the end of their life cycle. The exact percentage can be easily calculated, but inevitably there is a great amount of uncertainty in this classification for storms occurring during the early period of the records. A calculation using only storms from the satellite era, for which cloud structure and the change from the symmetrical pattern of a tropical cyclone to the comma-shaped pattern of an extratropical cyclone can be observed, gives a more accurate result. Quite a significant portion of all tropical cyclones, approximately 45 percent, undergo extratropical transition.[1] In 1938, however, the phenomenon was not common knowledge—it had never been directly observed, much less quantified. Most meteorologists had not given it any thought, and only a handful of typhoon experts had informally commented that it was likely that most of the storms transitioned after their recurvature, as reported in a 1922 *Monthly Weather Review* article.[2]

The moment when a system ceases to be tropical and becomes extratropical is not clearly defined, since different structural and energetic aspects of a storm can transition gradually, not necessarily at the same time and not always in the same manner. A purely tropical cyclone and a purely extratropical cyclone are perhaps easiest to distinguish, but a storm normally undergoes a gradual transition rather than suddenly switching to a new mode of operation. The loss of the symmetrical cloud structure is one of the easiest features to recognize on satellite images and it provides a very good way for forecasters to see that the transition is either occurring or has already occurred. It is also common to look into the "core" of the system for further insight into the status of the transition. In the middle tropospheric levels of a hurricane, the air in the eye is warmer than the surrounding air; hence, tropical cyclones are said to have a warm core. Contrastingly, extratropical cyclones have a cold core. Observing the cloud structure and determining the presence of either a warm core or cold core can thus indicate the status of the transition. The presence of dry air disrupting the clouds and rain around the eye (or former eye), further contributing to the observed asymmetry, can also aid in the determination. Deciding when to officially start classifying a storm as extratropical can be somewhat subjective and the methodology for doing so has varied throughout the years. At times, the extratropical designation was maintained for practical reasons, so that a system that had for the most part dissipated but was still producing potentially hazardous weather conditions could be further monitored. To standardize the usage of "extratropical" terminology, a more general term, "post tropical," is now used to refer to systems that no longer possess the characteristics of a tropical cyclone.[3] Along the spectrum of post-tropical systems, a leftover, weakened area of low pressure would be at one end, while a fully developed transitioned extratropical cyclone would be at the other. The new terminology was adopted in 2009 and put into operational use by the National Hurricane Center (NHC) during the 2010 hurricane season. The newly revised database, HURDAT2, now classifies post-tropical systems into "EX" (extratropical) or "LO" (remnant low).[4] After enough years pass using the new practice, it should become easier to determine the percentage of storms transitioning to extratropical with more certainty. The most likely series of events for the 1938 Hurricane, all occurring within the same day of landfall on September 21, was that the storm was a purely tropical cyclone at the beginning of the day, transitioned to extratropical throughout the afternoon, and finished the day as a semistationary (and post-tropical) remnant low.

When a hurricane approaches higher latitudes during and after its recurvature, it typically accelerates while embedded in a faster steering environment. At the same time, it also encounters environmental conditions that no longer allow its continuation as a hurricane (see Chapter 4). The surface of the ocean becomes colder, sometimes suddenly (e.g., as the hurricane clears the warm Gulf Stream current and encounters much colder North Atlantic waters) and vertical wind shear becomes much stronger (since the generally westerly winds that eventually embed the storm normally increase their strength with height). The hurricane might also come in contact with frontal boundaries and the flow associated with upper-air troughs. These interactions with surface and upper features in the middle latitudes bring the proximity of colder, drier air that can infiltrate the hurricane environment, inhibiting cloud and thunderstorm formation. It is easy, then, to assume that these newly hostile environmental conditions will cause the storm to weaken and eventually dissipate, making it impossible for the storm to maintain its intensity. This, of course, can and does happen, but if the timing and conditions are right, the storm will be able to tap into a new source of energy. Air temperature contrasts can be transformed into energy available for maintenance and intensification (see Chapter 6 for more about the favorable conditions for this energy transformation). The change in energy source is the key component of the transformation of a tropical to an extratropical cyclone; however, it is a shift that cannot be directly observed. The now-transitioning storm might still dissipate or it might maintain its intensity as its pattern of weakening is mitigated by the new source of energy. In some cases, a burst of reintensification might even occur. The 1938 Hurricane experienced its maximum intensity (for which our best estimate is 160 mph) well before it approached New England, while still latitudinally aligned with southern Florida (around 25°N) and as it was commencing its recurvature on September 19 (and through very early on the 20th)—before it initiated its extreme northward acceleration. The storm's intensity then diminished (150 mph by midday on the 20th and 140 mph toward the end of the day, while near latitude 30°N). However, though not intensifying, the storm remained the strength of a category 3—a major hurricane in the Saffir–Simpson Hurricane Wind Scale—until its landfall during the afternoon of the 21st. In this case, then, it is more accurate to say that the Hurricane's weakening was lessened by its extratropical transition.

So it is that as a storm moves northward it accelerates, weakens, loses its symmetry, and then gains a comma shape as the drier air becomes wrapped around from the west and the humid air wraps around from the east, with fronts developing in the same way as they would for an extratropical cyclone. The new cloud arrangement causes new asymmetry in the rainfall to develop as the storm progresses, with higher precipitation amounts normally accumulating on the left side of the track. Even before the extratropical transition is completed, as the storm travels faster and faster, the winds around it become increasingly asymmetric (as explained in Chapter 6). The strongest winds at this time are not normally of hurricane intensity, but the entire wind field expands and gale-force winds, which are more easily maintained, now extend farther out from the center. When, after the passage of the Hurricane, Charlie Pierce interpreted the additional data at his disposal—the same data with which he constructed the analyses that he published in his paper on the meteorology of the storm—he quickly determined that the devastating system had a frontal structure and many characteristics of what could be recognized as an extratropical cyclone (see Chapter 5 for more about Pierce and his analysis of the Hurricane).[5] He deduced that this unexpected transition had occurred before landfall. He even drew some frontal boundaries on his maps as the storm was still offshore. His description suggests his surprise at what he clearly saw in the observations:

> During the night of the 20th the hurricane passed through a transitory stage. It changed from a tropical to an extra-tropical storm—an extra-tropical storm, however, that maintained hurricane intensity. Like most other extra-tropical storms, this one had a definite system of fronts. On the face of it, this sounds like an absurd statement, because meteorologists think of a deep circular low as one that contains no fronts.

He then goes on to describe the large-scale environment, including the air masses and fronts that had now become part of the Hurricane environ-

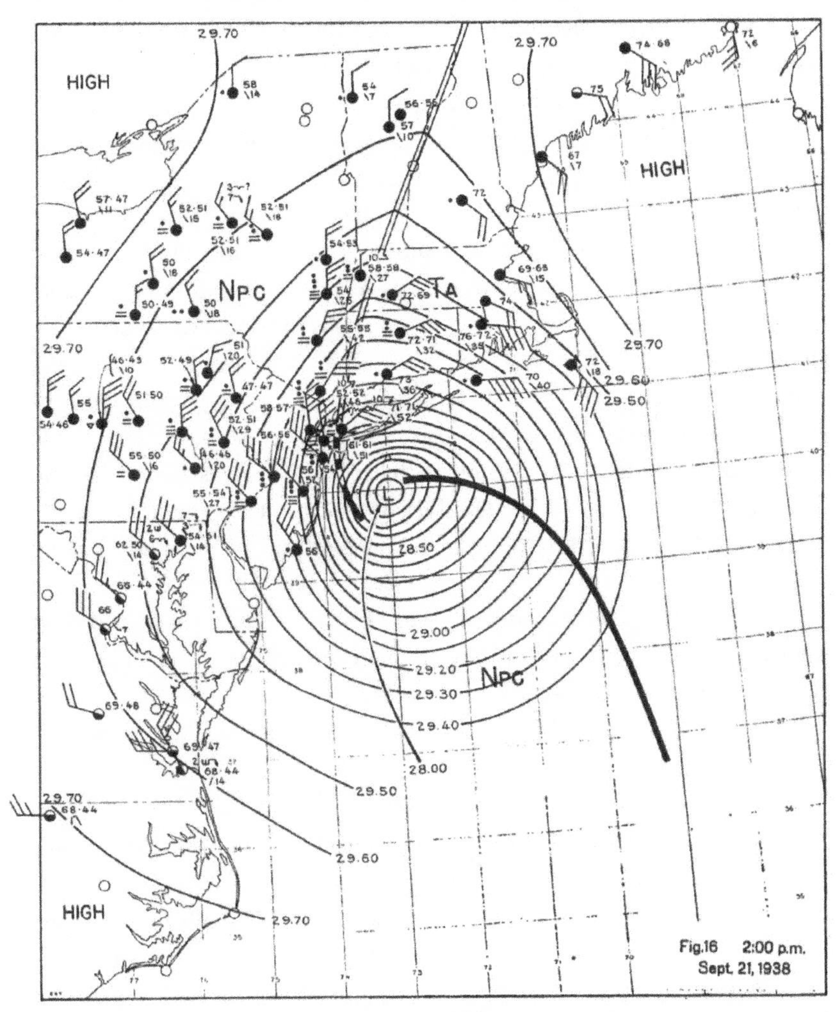

In his analysis of the storm, Pierce concluded that it had transitioned into an extratropical cyclone well before making landfall. This was the first time that such an observation had been made, although the possibility had been previously mentioned by typhoon experts. The 2 P.M. surface analysis from his paper (Pierce 1939) includes frontal boundaries (solid black for a cold front and white outlined for a warm front, as the convention of triangles and semicircles, respectively, did not yet exist). The current understanding is that this transition most likely did not happen until later, around the time of landfall.

ment and how they played into the storm's northward movement and its evolution into an extratropical storm. He additionally included a detailed explanation of the most likely change in the source of energy of the storm (as described in Chapter 6).

Since its creation, the original HURDAT identified the storm as extratropical during landfall—just as Pierce had determined it to be—but later studies contrastingly identified it as a landfalling hurricane.[6] More recently, the hurricane reanalysis project concluded that it is not possible with the available data to determine the exact timing of the transition, or rather of the completion of the transition to extratropical; the best that can be said is that the storm was most likely transitioning as it made landfall.[7] The revised HURDAT2, in which the 1938 reanalysis results have been incorporated, shows the storm landfalling as a hurricane and turning extratropical immediately afterward. The strong storm surge and winds experienced on the coast, together with the widespread inland effects, suggest that the description of "transitioning" is best. It is also the worst case scenario. If only the storm had fully transitioned before landfall, coastal impacts would not have been as severe. On the other hand, if the storm had never transitioned, the coastal effects would not have been followed by such widespread damage inland (the storm would have more quickly dissipated into rainy remnants). In other words, the relationship between the timing of landfall and extratropical transition in the Great New England Hurricane of 1938 maximized damages in intensity and scope.

Into the Forest

The mountains, rivers, streams, and trees whose abundance make northern New England's countryside so beautiful unfortunately also turned into weapons of great devastation in the presence of the Hurricane. The enlarged expanse of winds and precipitation resulting from the ongoing extratropical transformation in turn allowed more inland areas to be affected—a mountainous, forested terrain that was primed for disaster when the Hurricane quite literally struck the final blow.

Rain and More Rain

Popular and scientific historical accounts of the storm indicate that intense rains were experienced everywhere in the region during the days preceding the Hurricane. For example, the United States Geological Survey (USGS) report, "Hurricane Floods of September 1938," (described in more detail below) pointed out that "the hurricane as it struck New England was the climax of a 4-day period of rainfall which in itself was of outstanding amount and character. . . ."[8] The rainfall generally began with light showers during the afternoon of the 17th and continued essentially uninterrupted and increasing in intensity until the day of the Hurricane, when it briefly stopped right before its passage. To quantify this pre-Hurricane rainfall, the National Climatic Data Center (NCDC) historical weather data records for September 1938 once again help tell the story. Daily totals of precipitation, already used in Chapter 6 to calculate the total Hurricane rainfall, are recorded for every day for more than 150 villages, towns, and cities throughout New England. Most of these measurements were made by designated official observers in each town, and a few came from official Weather Bureau stations scattered through the region. The same intricacies described in the previous chapter about the timing of the observations also apply to the calculation of this preceding rainfall, except because we are now totaling four or five days (depending on the station), the introduced error is much smaller. Once again, for some of the stations, it is impossible to fully separate the last bit of the extended precipitation on the 21st (overnight or early morning) from the Hurricane rainfall (afternoon and evening). Additional rainfall data can be found in the USGS report, although some of the data in the report are not official Weather Bureau observations and hence do not appear in the NCDC records. Pre-Hurricane totals (September 17–20) show the largest amounts in Connecticut, Massachusetts, and the southern portions of New Hampshire and Vermont, ranging from four to more than ten inches. The largest total appears to have fallen in Barre, Massachusetts, where 15.36 inches fell during those four days, 11.83 inches on the 20th alone.[9] This observation does not appear in the NCDC data, for which the largest pre-Hurricane total is 14.17 inches, with 10.16 on the 20th for Hubbardston, Massachusetts.

Clearly, the amounts of pre-Hurricane precipitation were not trivial, and too large for a simple frontal system to produce on its own without the aid of a tropical moisture infusion brought northward by the circulation accompanying but still preceding the Hurricane. This phenomenon of intense prehurricane precipitation has been identified during the past decade as a relatively common occurrence and named a Predecessor Rainfall Event, or PRE.[10] As currently defined, a PRE comprises a well-defined area of heavy precipitation (observed to produce a rate of at least 4 inches in 24 hours) poleward of a tropical cyclone, clearly distinct and not associated with the storm's actual rain shield, but also clearly associated with the imminent arrival of the storm. Of course, there is no radar (to show precipitation) or satellite (to show cloudiness) data available for the 1938 Hurricane, making the corroboration of some of the listed features impossible. The entire precipitation event also spanned approximately four days. This period extends too far back before the arrival of the Hurricane for all of it to be directly connected to the storm, but locations with five or more total inches (some much higher, such as in Barre described above) just on the 20th were common. It doesn't seem possible for such massive amounts of rain to have occurred without the aid of the circulation preceding the Hurricane. There is no doubt about the major rain event, and even though we cannot verify it and not all of the details that we do know fit perfectly with the current definition of a PRE, at least a portion of the prehurricane precipitation was likely the result of such a feature.

As previously described, the Hurricane rainfall totals were significant but not extreme, ranging from approximately two to seven inches, but when adding the precipitation for both events (the preceding rain and the storm), some truly impressive numbers become apparent. The greatest totals (several of them above 15 inches) occurred along an axis extending from central Connecticut to central Massachusetts, but very significant totals (above 10 inches) extended farther outward from that axis and into southern New Hampshire and Vermont and western Long Island. The highest total rainfall for the period of September 17–21 was observed in two locations—neither of which appears in the official NCDC records—with more than 17 inches. As reported by the Connecticut State Forestry Department, Camp Buck in the Meshomasic State Forest received a total of 17.07 inches. Similarly, the town of Barre in central Massachusetts received 17.03 inches, as measured by the Massachusetts Department of Public Health. The highest total within the NCDC data was recorded in nearby Hubbardston with 15.6 inches.

Fallen precipitation is absorbed by the ground as much as possible within its current capacity. In New England, the soil in late summer and early fall normally has a large capacity for absorption, thanks to the roots of the abundant mature vegetation absorbing as much of the moisture as possible and high temperatures allowing large amounts of evaporation and transpiration from the plants. In other words, at this time of year the absorption capacity of the soil is quickly recharged and sporadic precipitation events can be processed easily. Even the rainfall accompanying a tropical system can be potentially handled by the soil without resulting in extreme flooding. A persistently wet season, however, would certainly diminish this absorption capacity by not providing enough opportunity for processing the water already absorbed by the terrain. As an example, both the recent Hurricane Irene (2011) and Hurricane Floyd (1999) dropped approximately five inches of rain throughout most of Vermont.[11] However, severe flooding accompanied Irene, which followed a very wet period, while very little flooding occurred during Floyd, which came after a remarkably dry summer.[12]

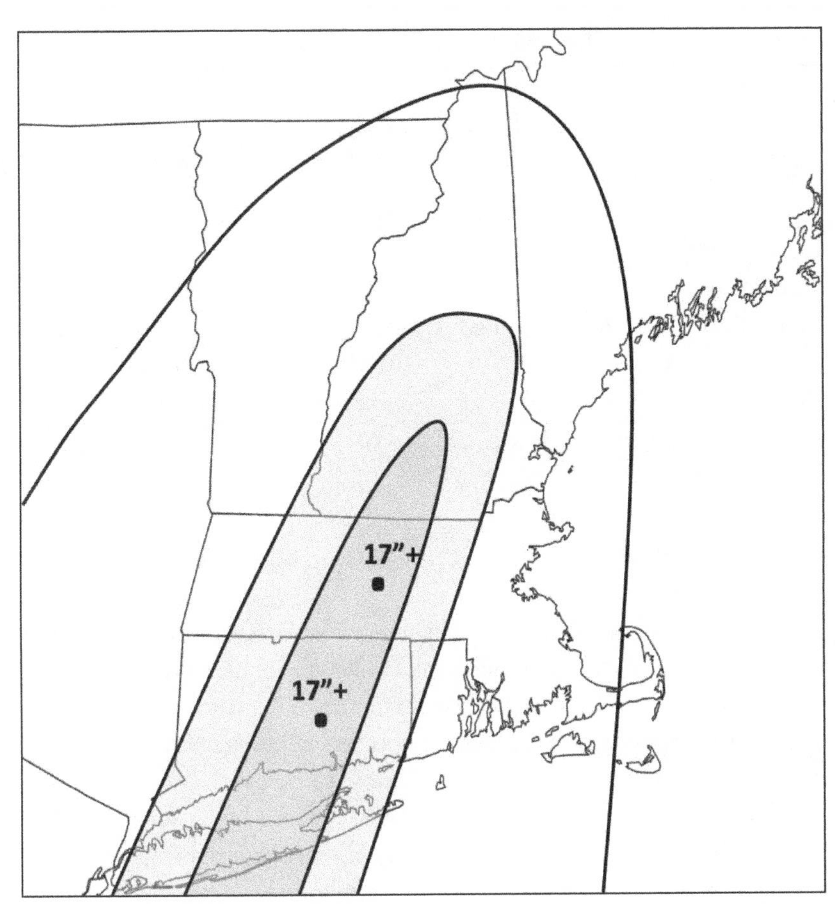

Significant precipitation was experienced throughout New England in association with the Hurricane together with its predecessor rainfall event, including an axis of truly massive amounts from Long Island, through Connecticut, central Massachusetts, and into south-central New Hampshire. The outside contour in this schematic representation is one inch of rain, the middle is five inches, and the highest is ten inches. Maximum amounts of 17-plus inches were measured at the Meshomasic State Forest (17.07") and Barre (17.03").

The 1938 Hurricane and its predecessor rainfall brought as much as three times that amount of rain, making it unlikely for the ground to be able to absorb and the rivers to be able to move all of it, even under normal soil moisture circumstances. In this case, the events also followed what had been a very wet season. Contemporary accounts of the storm all mention that the previous summer had been particularly cloudy and rainy. This suggests the likelihood that soil moisture was uncharacteristically high for that time of

TABLE 7.1. Summer 1938 Total and Departure from Normal Precipitation

	June		July		August		September 1–16	
	Total	Departure	Total	Departure	Total	Departure	Total	Departure
New Hampshire	4.00	+0.45	7.24	+3.56	3.64	+0.05	2.80	+0.94
Vermont	3.32	−0.30	6.84	+3.07	4.15	+0.55	2.86	+1.02
Massachusetts	7.39	+3.89	8.97	+5.45	3.21	−0.54	2.44	+0.53
Rhode Island	7.18	+4.10	4.40	+1.33	3.38	−0.31	2.05	+0.23
Connecticut	6.92	+3.44	9.57	+5.61	4.20	+0.03	1.90	−0.17
Eastern New York	3.97	+0.10	5.49	+1.74	4.68	+0.61	2.51	+0.01
New Jersey	7.79	+4.08	8.84	+4.05	3.09	−1.66	1.02	−1.06

SOURCE: Data obtained from USGS (1940) Water-Supply Paper: "Hurricane Floods of September 1938"

year. It is desirable to quantify this claim of an unusually wet summer by determining how much rain had indeed fallen compared to normal amounts. By averaging daily 1938 rainfall at all stations in each state and comparing to the "normal" rainfall for each month (which in this case is the average for the previous 51 years, starting at the beginning of the data record for most of the stations in 1886), it is easy to verify that the entire region had received much more rain than normal.[13,14] Four or five surplus inches were the norm for all of New England during the summer (June, July and August, or JJA) and the first half of September, with the southern region being especially wet (see Table 7.1). New Hampshire, for example, received exactly five inches above normal, corresponding to 128% of its normal precipitation. Massachusetts and Connecticut had the wettest summers with 9.33 (142.4%) and 8.91 (139.4%) excess inches, respectively.[15] These numbers are statewide averages, but some individual stations had much higher anomalous totals, making corresponding localized spots tremendously wet. The Blue Hill Observatory, for example, had more than doubled its normal JJA precipitation of 12.01 inches, with a total of 24.03, by the beginning of September.[16]

Flooded Rivers

Even before the excessive rainfall during the days preceding the Hurricane, the ground already held a large amount of moisture. The four days of increasing rainfall then caused the excess water to run off the many mountainous slopes and, before the arrival of the storm, most bodies of water were well on their way to being or were already above flood state. The Hurricane then arrived with a significant amount of additional precipitation, nowhere near extreme for a tropical cyclone, but still enough to provide the extra water to guarantee imminent havoc.

Unfortunately, the region had already suffered similar flooding events during the past decade and a half, as some of the worst floods in New England history occurred within the same two decades. Eleven years before, in November 1927, the remnants of a late-season tropical storm stalled over the region, severely flooding Vermont, New Hampshire, and parts of Massachusetts and Connecticut. Only two years before the Hurricane, in March 1936, two intense rainstorms followed an especially cold and snowy winter, causing widespread record-breaking flooding throughout all of New England from the combination of the snow melting, ice jamming along the rivers preventing their flow, and the rainfall itself.[17] The floods accompanying the 1938 Hurricane surpassed in some locations or at least challenged the record flood levels from these two previous events at many river basins (Merrimack, Hudson, Thames, Connecticut, Housatonic, and Raritan Rivers and Lake Champlain) and their tributaries.[18] Even today, these three historical flood events either individually or in some combination appear in the list of top crests for many rivers in the region. In Plymouth, New Hampshire, for example, the three highest flood levels for the Pemigewasset River (with a flood stage of 13 feet and major flood stage of 21 feet) are 29.00 feet on March 19, 1936; 27.40 feet on November 4, 1927; and 23.62 feet on September 21, 1938.[19] Much more recently, the flooding accompanying Hurricane Irene also joined the list at various locations. In Plymouth, Irene produced the seventh-highest river height, 21.69 feet, on August 29, 2011.[20]

On Wednesday morning, September 21, rivers were full to their banks. As the Hurricane moved through the area, floodwaters continued draining down the heavily forested slopes of the Berkshire, White, Green, Adirondack, and Catskill Mountains and moving toward the valleys and plains. River and stream water levels, which had uniformly and gradually started to increase early in the day on the 20th, continued rising during the 21st but now quite sharply. The flow of water far exceeded the capacity of most river channels. As the engorged bodies of water deepened and widened, adjacent lands were completely inundated. Bridges and dams were destroyed or damaged and nearby buildings were swept away. Everything in the flood plains was ultimately ruined to various degrees by the water and its force.

Plentiful data are available to quantify the magnitude of the inland flooding. The height reached by the water during a flood, typically measured in feet from the bottom of the river (but not exclusively, as other reference points can also be used), is termed "gage height" and it is more commonly

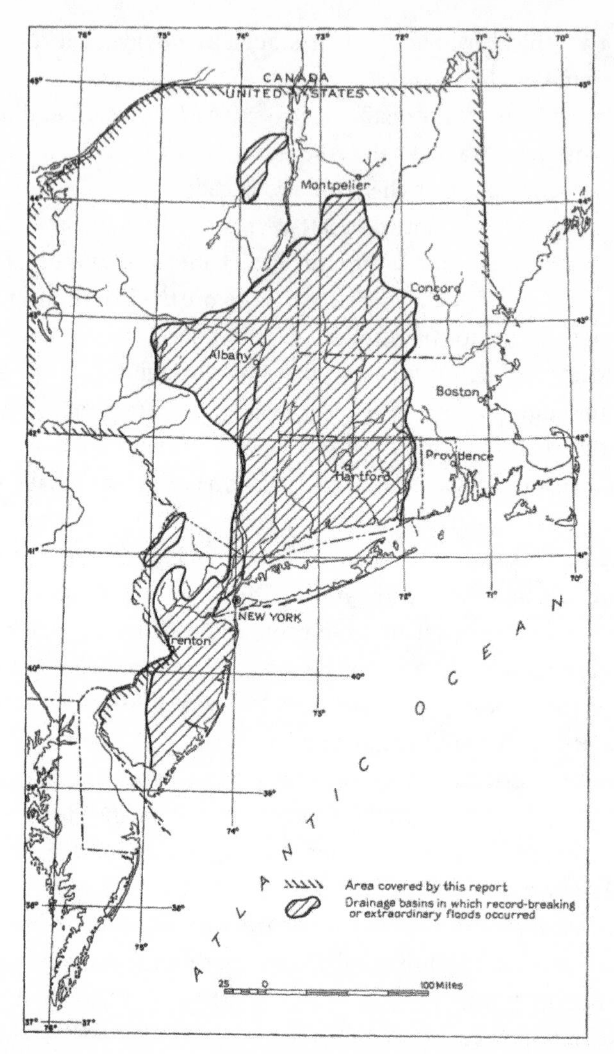

This map from the USGS (1940) report on the "Hurricane Floods of September 1938" shows the areas affected by the worst floods (hatched region). The detailed study of all relevant hydrological aspects covered New Hampshire, Vermont, eastern New York and Long Island, Massachusetts, Connecticut, Rhode Island, and New Jersey.

used to communicate flood levels to the general public. Hydrologists instead prefer to use a water flow measurement known as discharge, a less tangible quantity but more representative of the amount of water being moved by a river. Discharge is technically defined as the rate at which a volume of water is transported through a river cross section, and it has traditionally been measured in cubic feet per second, or cfs. The USGS conducted a massive

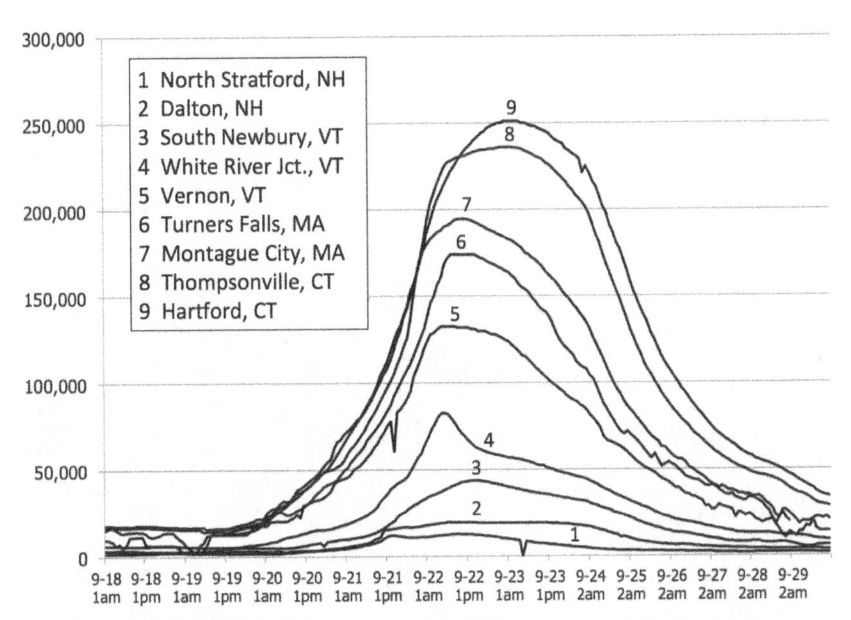

River flow measurements at various USGS stations along the Connecticut River and its tributaries from September 18–30 show that levels started to rise on the 19th and 20th and peaked within a day or two of the Hurricane's landfall, with higher and later peaks at downstream stations. The flow is measured in cfs (vertical graph axis).

study of all hydrological aspects of the Hurricane floods, published two years later as a 500-page report (containing hourly precipitation at some select locations, soil moisture, absorption and storage, run-off studies, and more importantly for this study, river discharge and gage heights, all organized in tables, charts, maps, and detailed descriptive analyses).[21] A large portion of the data in the USGS report consists of discharge values in cfs units (at the time referred to as "second-feet") for more than 300 river stations. Using just a fraction of this data reveals a relatively clear picture of how the inland flooding progressed.

In general, waters within the southern portions of individual river basins started to rise first; the northern portions then also rose but peaked faster and at lower flow and height levels. The later and higher crests farther south are easy to understand, since it takes time for water to move downstream—and the farther downstream, the greater the volume of water that is added to the river from the surrounding slopes and tributaries. The result is higher peaks downstream, sometimes much higher, depending on the size of each specific basin. Some locations peaked just as the precipitation from the Hurricane was ending, while others peaked during the next day or so. By the

The Music Shell at Bushnell Park in Hartford, Connecticut, reflecting on the calm flooded waters on September 22 is one of the most iconic pictures of the storm. (NOAA Photo Library)

23rd, all or most of the waters were already subsiding, although it would take a few days to reach levels similar to those during the middle of September, before the predecessor precipitation event started.

The Pemigewasset River in central New Hampshire is a tributary of the Merrimack River, which crosses into Massachusetts where it turns east and eventually discharges the waters gathered deep within the White Mountains and throughout the Lakes Region straight into the Atlantic Ocean near the border between the two states. The USGS has maintained a gaging station in the town of Plymouth, roughly halfway along the Pemi (as it is known to the locals) since 1903.[22] A close look at the data recorded at this station provides a better picture of the hydrological events in northern New England on that day.

Even though the preceding months had brought wet conditions throughout the area, the Pemi was very low, just about 1.5 feet, as is normal for early September. The water flow was about 1,000 cfs. Light rainfall during the evening of the 17th increased to moderate intensity during the 18th, and heavy

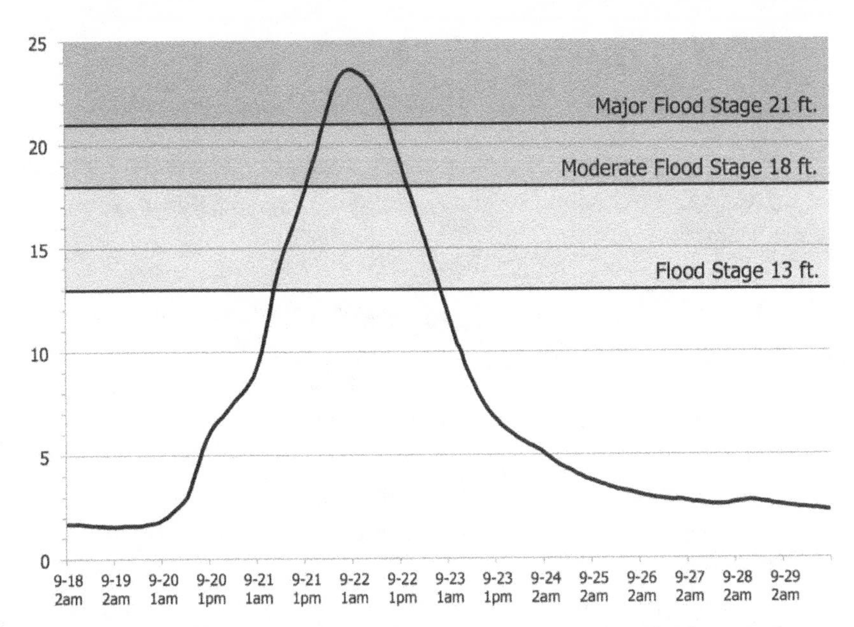

The Pemigewasset River began rising early on September 20, reaching flood stage early on the 21st, further and quickly rising to major flood stage level throughout the day. It took a couple of days for the water levels to go back below flood stage and a few more to get back to levels similar to before the storm and the preceding rain event. Water height as measured at the USGS station in Plymouth, New Hampshire (USGS 1940).

rain then dumped more than an inch of water on September 19. The river finally started to rise early on September 20, right before more than three and a half additional inches of rain fell. A combination of the extra intense rainfall and the excess waters making their way downstream from the White Mountains caused the river levels to increase steadily throughout the day. By midnight, gage height was up to 9 feet—a large daily increase to be sure, but still well within the capacity of the Pemi. By 5 A.M. minor flooding had started in the river valley lowlands, as it officially reached its flood stage of 13 feet, and by 9 A.M. the waters had reached the road connecting Plymouth and Holderness (now state highway 175A). The rapid increase continued and moderate flooding levels had commenced as of about 2 P.M. At this time, rainfall from the Hurricane, which was just making landfall more than two hundred miles to the south, was most likely starting to fall over the region. By 6 P.M. major flooding levels of 21 feet had been reached, and the river now covered the road with about four feet of water. The crest of 23.62 feet was not reached until 1 A.M. on the 22nd, well after rainfall had ended over central and northern New Hampshire.

FLOODING CATEGORIES

The severity of a flood at a specific location is defined in terms of the impact of specific water levels at the location. The flood categories used by the NWS are as follows:*

- Minor Flooding—minimal or no property damage, possibly some public threat
- Moderate Flooding—some inundation of structures and roads near stream, some evacuation of people and/or transfer of property to higher elevations
- Major Flooding—extensive inundation of structures and roads, significant evacuations of people and/or transfer of property to higher elevations
- Record Flooding—flooding level equals or exceeds the highest state or discharge at a given site during the period of record keeping

* NWS Glossary.

The waters took longer to subside than they took to rise: at 10 A.M. the next morning they finally went below major flooding-category levels, and at 7 P.M. the water finally vacated the road. At 11 P.M. levels went back below 13 feet and, even though the river channel was still completely full, the water was now well contained and thus no longer causing trouble for anyone in the vicinity, where cleanup activities had likely already begun. Water levels continued to decrease gradually throughout the next few days and finally started to stabilize a couple of feet above the levels before mid-September by the 25th and 26th. The highest discharge during the entire event was 34,800 cfs (falling well short of the record levels reached during both of the aforementioned preceding floods in 1936, with 57,300 cfs, and 1927, with 40,100 cfs), and by the end of the week, levels appeared to be stabilizing around 1,500 cfs, somewhat higher than pre-Hurricane levels.[23] All buildings within the flood plain were inundated by the waters of the normally peaceful Pemi. First-hand observation of the conditions during Hurricane Irene, which caused a crest of 21.69 feet, gives more recent perspective on what the conditions were like in 1938. The waters were probably raging and turbulent during the evening of the 21st, then calmed down to seemingly quiet but still-moving waters by the morning of the 22nd, with sporadic debris—a tree, a chair—floating down the main river channel as onlookers stared at the remaining flood. Finally, after the waters receded, damaged furniture, moldy basements, and brown sediment remained to be cleaned up.

The soft, saturated ground brought about a variety of problems. The raging water in the engorged streams and rivers scooped and carried away ample amounts of soil, causing great turbidity and leaving behind permanent erosion. The rivers, however, were not the only movers of the loose, wet soil; gravity and the mountainous terrain also did their part. Numerous landslides were reported, especially within the White Mountains. Without question, however, the most dramatic aftermath resulted from the ground losing its ability to firmly anchor tree roots. The Hurricane still carried significant winds, especially in locations where the topography served to enhance them. It was easy, then, for these winds to knock down the uncharacteristically vulnerable trees.

Decimated Forest

Millions and millions of trees were broken, bent, and uprooted by the Hurricane, which left behind a tangled and chaotic landscape unimaginable to modern New Englanders. Blow-down surveys were immediately attempted by land owners and other interested parties to determine the potential to salvage the wood to offset the impending catastrophic economic losses. This was not an easy task, considering that many areas were impenetrable, in addition to the uneven character of the damage and the lack of reliable prestorm survey data for large portions of the area. The ensuing estimates immediately following the storm varied considerably, ranging from as "low" as 1.5 to as high as 5 billion board feet of timber—a common forestry measurement for trees, logs, and lumber.[24] One board foot is the volume contained within 144 cubic inches of wood, and it is most commonly identified as a one-inch-thick, 12-inch-square block of wood (12"×12"×1"). Once damage was assessed more adequately throughout the following months, the total estimate provided in reports from all New England state foresters was 3.6 billion board feet, and the final estimate by the U.S. Forest Service was somewhat lower, closer to 2.6 billion.[25] Because timber salvage was the motivation behind the estimates, they did not include trees traditionally not available or desirable for logging, such as ornamental and shade trees in public parks and recreation areas, town commons and streets, or state forests and parks. Young forest trees of no immediate logging value were also not counted. The total tree loss was, therefore, even higher than suggested by the reported estimates. Calculating the number of trees represented by the board feet estimates can be somewhat tricky. One could choose an average tree height and width, calculate the corresponding volume, and divide by 144 to determine the number of board feet contained in a hypothetical tree, and then calculate how many times this quantity was contained in the estimated board feet loss. This would result in a very rough estimate of the number of downed trees. For a 50-foot-high, 10-inch-wide tree, this simple calculation translates into approximately 3 million trees for each billion board feet, for a total of approximately 9 million trees.[26] This is in all likelihood a great underestimate, as trees are not perfect cylinders and their width obviously tapers with height. To make matters more complicated, the relevant width and height varies immensely and the amount of available wood also depends on the grade and type of tree, making our estimate of 9 million highly inadequate. Additionally, the way in which board feet are calculated has much less to do with the volume of the tree than with an estimate of the usable amount of lumber contained in a log (making allowances for the "waste"), which in turn depends on the type and other specific characteristics of the tree.

A variety of well-established rules can instead be used to estimate the board-foot volume for any specific tree.[27] A brief examination of tables constructed using these rules easily shows that the number of trees blown down by the storm was more likely of the order of tens of millions, an order of magnitude higher than the rough calculation done using the volume of a cylinder.[28] These tens of millions of trees blocked roads and forest trails, caused structural damage to homes and other property, tangled the forest, and represented a significant fire hazard. The timber salvage operation that ensued was the largest in U.S. history (see Chapter 8).

The forested landscapes of all New England states were affected, some more seriously than others. Maine experienced the least damage and New Hampshire the most, with Massachusetts a close second. Approximately 35 percent of the region, or 15 million acres, was covered with downed trees, but the damage was not homogeneous.[29] Completely decimated, spottily damaged, and barely touched areas coexisted throughout the region. Exposed areas experienced more severe damage, and many forested mountaintops were left bald. The White Mountain National Forest (established in 1918) in north-central New Hampshire was especially devastated. One thousand miles of trails were blocked, and the roads not washed out or blocked by landslides were "crisscrossed by as many as 200 trees per mile." The tree

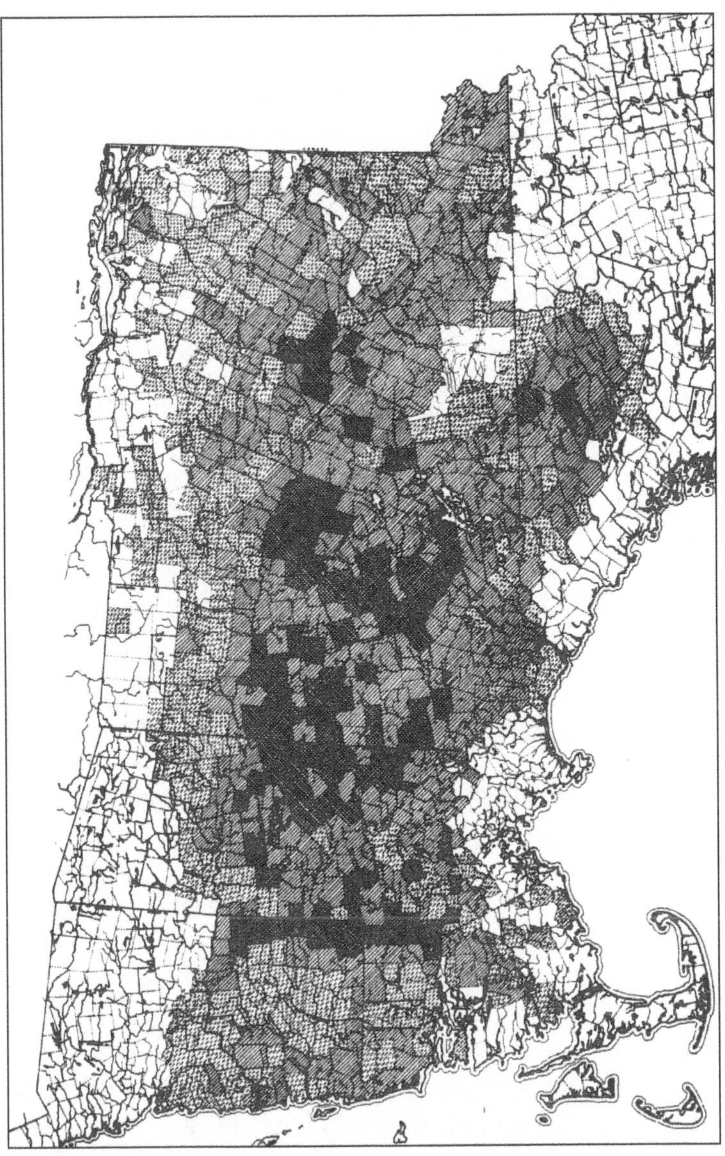

Immediately after the Hurricane, the Forest Service conducted a massive survey of tree damage throughout New England as a foundation for the fire hazard reduction and timber salvage programs. The former was considered of more urgency for public safety and the latter was a massive, unprecedented endeavor for which the New England Timber Salvage Administration (NSTA) was especially created. The map shows estimated tree damage categorized as fire hazard (darkest shading is "extreme fire hazard," or over 10 million board feet down [bfd], the next is "moderate fire hazard," or one to 10 million bfd, and the lightest shading represents less than one million bfd per unit). Units correspond to firefighting districts, which in most cases match town lines (Forest Service 1943).

damage, as one might expect, was not exclusive to northern New England but also occurred throughout the southern New England region including the coast, where the salty sea spray and floodwaters from the battering waves and the storm surge caused additional vegetation damage.

The damage pattern in the face of a catastrophic wind event such as the 1938 Hurricane depends, as implied above, on site exposure (which itself depends on the terrain's slope angle and its orientation with respect to the wind), local soil characteristics, and vegetation height and type.[30] All other conditions being equal, taller trees, especially those 75 feet or higher, normally experience more damage than their shorter counterparts.[31] In terms of composition or type dependence, conifers are much more susceptible to what ecologists call windthrow, the breakage or uprooting of trees by the wind, in comparison to hardwoods. Indeed, during the passage of the 1938 Hurricane, pine trees (specifically white pines or *Pinus strobus*) were certainly the most disproportionally damaged. The New Hampshire state forester, for example, estimated that 75 percent of the downed timber was white pine—1.2 billion board feet of pine versus 0.2 of hardwoods and 0.2 of "other trees."[32] Hardwoods were a significant portion of the blowdown only in Vermont and Connecticut. Losses in Massachusetts and Maine were also mostly white pine. This transitional species had claimed large portions of the terrain formerly covered by agricultural fields. Deforestation in New England had increased throughout the 18th century and peaked during the early to mid-18th century when more than 80 percent of the region was open land.[33] Many fields had been abandoned since then and left to natural reforestation, resulting in extensive areas populated by white pines. These fast-growing trees not only covered a great portion of the affected terrain, but when they did share an area with other types of trees, they easily towered over even large hardwoods. Near Flume Gorge (the Flume) in the White Mountains, the then-famous "Sentinel Pine" tree stood guard for a hundred years, maybe much longer, on a high cliff above the Pool, a deep basin in the Pemigewasset River (farther upstream than the Plymouth USGS station). This respectable specimen, at least 100 feet tall and 5 feet wide (with a circumference of 16 feet), survived the massive deforestation of the previous centuries, but it did not survive the Hurricane.[34] Volunteers from the Society for the Protection of New Hampshire Forests cut a 60-foot-long plank from the fallen giant and placed it spanning the rocky cliff walls of the gorge, high above the stream leading to the Pool. A footbridge was then quickly constructed using additional wood from other fallen trees in the surround-

The Sentinel Pine Covered Bridge now stands near where the giant tree once stood watchful over the surrounding forest near the Flume Gorge in the White Mountains. A 60-foot plank cut from the fallen tree now serves as the support beam for the picturesque footbridge.

ings and the Sentinel Pine became the support beam for a classic covered bridge.[35] The bridge stands today as a reminder of the devastating storm in a landscape that appears otherwise unscathed.

Beyond the complete desolation brought to coastal areas by the storm surge, to those who experienced the storm and its aftermath, the severe loss of trees was one of the most shocking sights. A New Hampshire newspaper poetically declared: "Our foliage friends have been slain and many hearts are sad . . ." but went on to acknowledge that "yet our uncounted losses are meager compared with the loss of life and destruction in Rhode Island and many other places. We have much to be thankful for."[36]

A wonderful and poignant collection of memories from the storm is contained in a sort of scrapbook put together during the months immediately following the storm by a Connecticut school district.[37] Teachers collected drawings, writings, poems, and even songs composed by the children. Some were prompted and some were created spontaneously during the six months following the Hurricane. Even though southern New England did not experience the most severe tree loss overall, it did suffer the greatest loss of shade

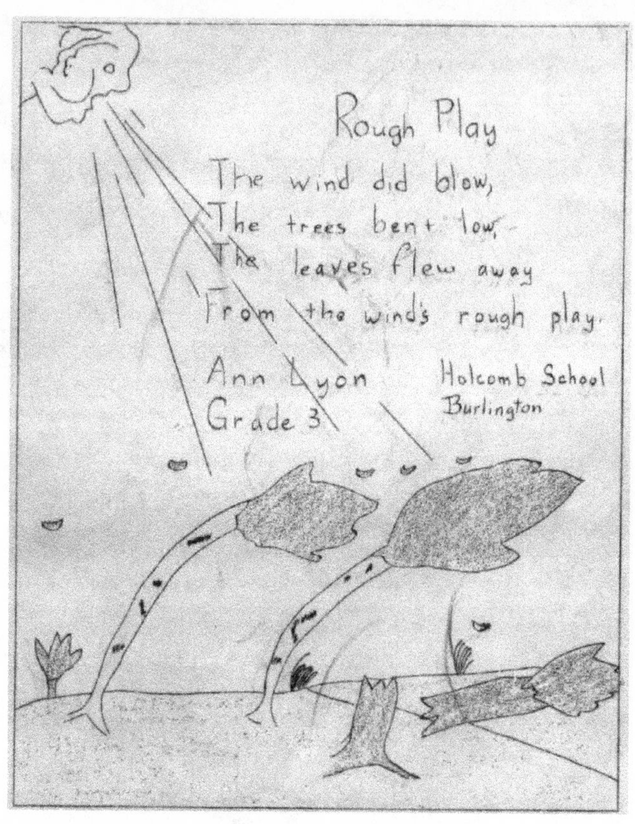

A scrapbook put together by the Connecticut school district #7 contains hundreds of stories, poems, songs, and drawings by local schoolchildren. What happened to the trees was prominently featured in many of the pieces. This drawing and poem came from a third grader in Burlington, Connecticut. (Girls and Boys of the Connecticut Supervisory District #7)

trees in town commons and streets. What happened to the trees seems to have made quite an impression on the little ones' minds:

> How the trees bent down to the ground!
> They swung around and 'round
> How the wind through trees did moan
> It made the oaks and maples groan.
> (Carl Royko, Grade 4)

> The wind was oh so very bold,
> The earth the trees just could not hold.
> (Doris Deming, Grade 4)

Many other poems and songs and general references to the trees in addition to various drawings can be found throughout the more than 80 pages.

The Stormy End

What was left of the Hurricane brought a blustery, stormy evening to interior northern New England. It is not hard to find brief local accounts of the Hurricane at small libraries and historical societies throughout towns and villages in and around the White Mountains. They all speak of the wind, the rain, the washed-out roads, and the impenetrable obstacle course of fallen trees and tangled branches. Most people were taken by surprise; without a full grasp of what was going on, they intuitively reacted to avoid dangers and to protect their loved ones, themselves, and their property. Once it was over, the unrecognizable landscape was a heartbreaking sight.

After the storm made landfall in southern New England, the winds very quickly diminished from the strength of a major hurricane to that of a minimal hurricane, and then further weakened to less destructive but still significant tropical-storm strength (40- to 70-mph sustained winds, but also stronger gusts aided by the channeling and turbulence provided by the mountainous terrain). As the worst part of the storm was making its way through New Hampshire and Vermont, it maintained this gale strength, still quite sufficient to cause widespread damage. In northern New England, the Great Hurricane would climax this evening:

> . . . The atmosphere shed a peculiar glare about three o'clock in the afternoon. Then the wind began to blow and people decided to fasten awnings and remove their summer furniture to safety. By six o'clock, limbs begun to break off the shade trees while the wind increased to a roar. All night the hurricane raged, damaging houses, barns and forests with terrific velocity.[38]

> All our people remember well the weird wild light in the murky sky, the howling shriek and roar of the gale. Scarcely heard was the crashing all around of thousands of trees but, as they fell on wire lines, telephones and lights went off. Soon roads were blocked by trees and floods, as higher and higher the rivers rose, fed by mountain brooks grown into raging torrents. Evening came and slowly the fury of the blasts died down.[39]

The storm was finally over for northern New England. Pierce did not track the storm beyond midnight of the 21st, but HURDAT/2 keeps it around for a little longer, still spinning and raining into Quebec and perhaps the

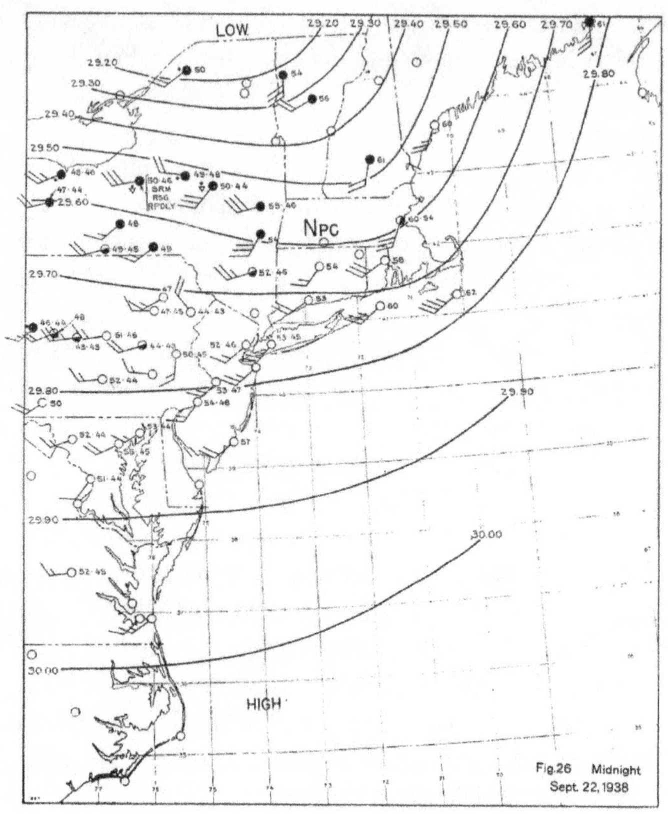

The last surface map drawn by Pierce (1939) to wrap up the analysis of the passage of the Hurricane was the midnight map for September 22. A large area of low pressure can be noticed on the top left portion of the map, now well off New England.

most northern areas of New England. Well past extratropical transition, this was not a tropical hurricane any longer, and it was barely an extratropical cyclone. The modern post-tropical terminology would have applied very well in this case.

The Daily Weather Maps (see Chapter 2) for the next handful of days after the Hurricane's landfall offer some distinct clues as to the end of the storm's journey. September 22 clearly shows a still-strong low pressure system that had not moved much farther northward than the previous day, centered over Quebec. Additional rain and winds would have still accompanied the system at this time. The lack of forward progress means that the storm had most likely occluded at this time. As an extratropical cyclone nears the end of its life cycle, the basic source of its energy—the contrast between the moist

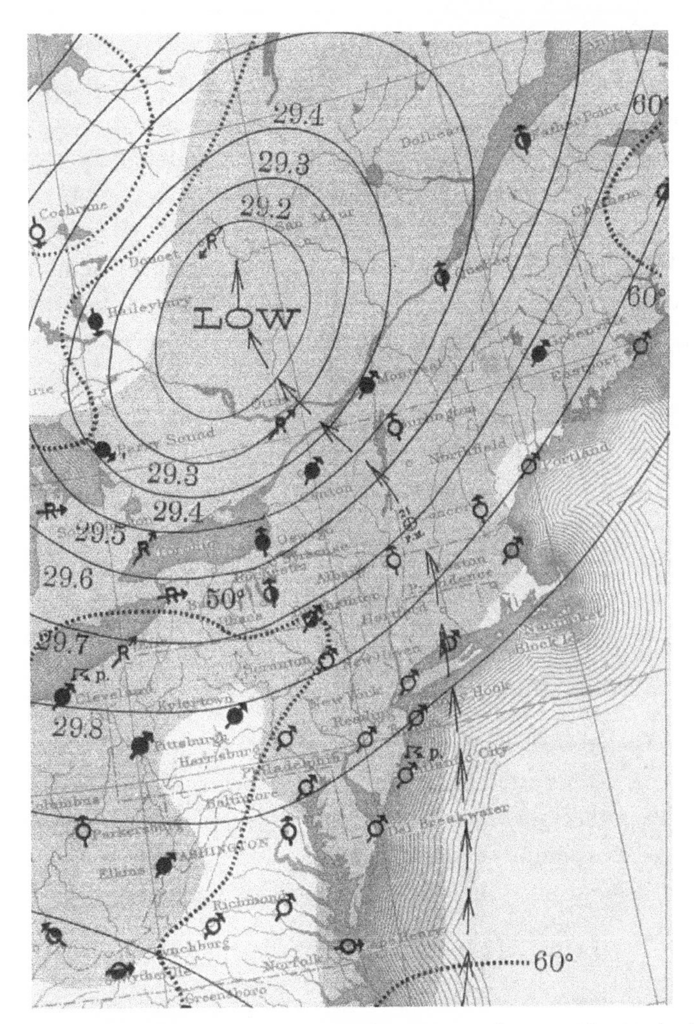

A portion of the Daily Weather Map for September 22 zoomed-in over New England clearly shows the remnant low and previous day's precipitation (NOAA Central Library).

tropical air and the cold polar air—is cut off. The cold front catches up to the warm front and the new combined front (known as an occluded front) now wraps around the storm. At the same time, at this stage, the storm dislodges itself from the jet-stream steering currents that have been moving it and just spins itself down. September 23 shows a weaker low pressure area in the same general region. The remnant low of the 1938 Hurricane spun in place, dissipating and being replaced by the weather systems coming from the west. September 24 and 25 show high pressure now generally taking over the region. The clear, sunny weather that accompanied this area of high pressure was indeed good news for the relief and rescue efforts that commenced immediately after the winds died down and the rains stopped.

The Hurricane was no more, the extratropical cyclone was no more, and it was time for New England to start recovering:

> . . . When morning dawned, the havoc was too astonishing. The earth was so soaked that century old elms and maples fell, pulling their roots from the ground. Every street was blocked by its shade trees. The elms on the east side of the Village Green lay prostrate. . . . The forests were a tangle of broken, uprooted and topless logs. . . . While this seemed unbelievable, a salt spray covered everything, mingled with finely ground leaves that dried to a crust that demanded scrubbing to remove from walls and windows of the houses. Electric wires were tangled among the fallen poles; telephones were silenced for days. . . .[40]

But it would take more than just days, rather years and even decades before the scars left behind by the Hurricane started to blend into the landscape.

The storm had traveled more than 5,000 miles throughout its lifetime. In just one day, the Great New England Hurricane of 1938 had covered a distance of 1,000 miles and left behind unimaginable destruction of life, property, and landscape. As its remnants spun themselves out over southeastern Canada, the Hurricane also left behind a complex historical and meteorological legacy.

PART III
AFTERMATH

8

WHAT WAS LEFT BEHIND

Storm impacts and the relief efforts

A great swath of devastation now spanned from the beaches of Long Island to the mountains of northern New England. In less than half a day, the force of the ocean water, the wind, and the excess rainfall had flooded, swept, eroded, broken, and knocked down natural and hand-built structures. The ocean quickly retreated and the winds and rain ceased during the evening; the rivers, on the other hand, would still rage for a while longer. The amount of water flowing through the rivers and streams across the region was just peaking or still increasing when the stormy weather was ending. At some locations, especially toward the downstream end of the largest rivers such as the Connecticut and the Merrimack, it wasn't until a day or two later that the water reached its maximum height. Consequently, as the tree and debris cleanup commenced, some towns still had to deal with additional or impending river flooding. It was fortunate, then, that a period of clear weather came soon after the Hurricane passed. The massive amounts of rain during the previous week had topped that of the preceding rainy months, but the weather now provided favorable conditions for the huge job of rescue, relief, cleanup, and reconstruction faced by the entire region.

Forecasts and surface weather patterns from historical Daily Weather Maps archived by the National Oceanic and Atmospheric Administration (NOAA), in conjunction with the corresponding daily precipitation amounts and temperatures from archived National Climatic Data Center (NCDC) daily data, reveal exactly what kind of weather dominated New England during the days following the Hurricane. After the exceedingly rainy period during the middle of September (that culminated with the Hurricane), the weather indeed cleared. A more normal fall weather pattern of rain-producing systems passed through every three to seven days (instead of heavy rain that extended through several consecutive days). From a total of six or so events during the next month and a half (all of which consisted of extratropical low-pressure systems and their accompanying fronts), only one toward the end of October produced significant rainfall. This system, which was clearly extratropical by the time it affected New England on October 24, was originally thought to have been a tropical storm while over the Gulf of Mexico. The hurricane reanalysis project, however, just recently determined that the system was most likely extratropical throughout its entire lifetime.[1] Regardless of its origin, it brought one to two inches of rain throughout the entire region. Comfortable temperatures (highs in the 50- to 70-degree range), dry weather (with dew points mostly in the 40s based on the overnight low temperatures), and varying degrees of cloudiness dominated the days in between these brief rainfall events, with the longest stretch of such conditions occurring during the second week of October.

At a time when New England would normally be expecting the vibrant reds, oranges, and yellow leaves of autumn, its forested landscape had instead violently transitioned, at least visually, from lush summer to barren winter. Sunny blue skies, mild temperatures, and low humidity provided a contrasting backdrop to the unbelievable scene of destruction, with broken structures and bare, fallen trees scattered everywhere.

Aftermath, Revisited . . .

The damage was absolute, even though localized rather than general, along portions of the Long Island and southern New England coasts. Coastal areas that were spared the overflowing ocean and major hurricane-wind intensity, as well as interior New England, were unevenly affected. The damage was spotty yet serious and widespread. The scene of devastation left behind is hard to imagine for those who have never lived through such an experience. Reports, articles, personal accounts, and photo collections paint a shocking picture, but against this overwhelming backdrop an uplifting portrait of survival and resilience also emerges.

Hundreds of small and large cottages, pavilions, and "amusement structures" once dotted the low-lying beaches that outline the region.[2] By the time the Hurricane had come and gone, very few were still standing. Those

A house blown onto railroad tracks by the Hurricane of 1938. (Courtesy of the Boston Public Library, Leslie Jones Collection)

buildings that did survive the battering of the ocean were mostly empty, uninhabitable shells. Entire houses or large portions of houses that did not initially break down into pieces of lumber were taken off their foundations, stripped of their furnishings, and carried inland by the water. Boats of all types and sizes were destroyed as they were smashed against rocks and seawalls or dragged inland over roads, backyards, and graveyards. Everything was covered in sand and seaweed.

Before-and-after photograph displays, popular in the write-ups and photo collections published soon after the storm, show the destructive power of the Hurricane. Beaches once lined with piers, homes and other structures were now seemingly naked. Of course the debris all had to go somewhere, and a huge expanse behind the beaches was covered with chairs, beds, refrigerators, boats, and lumber deposited at elevations as high as 5 to 15 feet above the mean high tide (demonstrating how much higher than normal the water reached).[3] The debris field stretched for miles along the inland side of the shores, marshes, and ponds.

Much of the loss of life caused by the storm occurred in these devastated portions of Long Island and southern New England. The small state of Rhode

Island, unfortunately located just to the right of the eyewall and possessing topography particularly suitable to enhance the already maximized Hurricane storm surge, suffered hundreds of fatalities, the highest among all affected states.[4] The unexpected arrival of the storm meant that townspeople were going about their business as the violent weather began. Everyone would have noticed the start of the rain and the wind, but at first would have had no reason to believe that anything different from ordinary stormy weather was upon them. Many were out and about on the streets, with the wind blowing their umbrellas inside-out, perhaps causing some excitement and motivation to run to the next shopping or errand destination. The excitement, however, soon would have turned to horror at the advance of the ocean.

While many lost their lives, a much larger number survived their ordeals, many of them in dramatic fashion. Some people were at work when the ocean surged onto land, the winds started blowing, and the trees began falling. Many were forced to react to the changing elements, rushing to save equipment or even their own or others' lives as the water rushed through the windows. Others were at home or picnicking on the beach, where they would have been helpless against the advancing ocean. Still others made their way to safer homes or higher ground, or managed only to get on a boat or a piece of floating wreckage. The chaotic waters tossed both people and debris about, and when the ocean finally retreated, it revealed a grim scene. Bodies floated in shallow water, laid on the beach, or were left buried under the sand or tangled in the wreckage.[5] The exact number of those killed by the storm might never be known, but the numbers reported by different sources roughly range from 500 to 700.[6]

Clearly, only the immediate coast was affected by the storm surge, but a much larger coastal area experienced the strongest winds accompanying the Hurricane. Within the southernmost portions of the region, the roar of the hundred-plus-mile-an-hour winds would have been deafening. Children were still in school at this time, and many were dispatched early, sent home in buses, or picked up by parents. Some children were simply allowed to walk home on their own, but many others were kept in school and then let out once the danger was perceived to be over. An account from the high school in the town of Bridgehampton on the eastern half of Long Island suggests that in some cases this was perhaps done a little prematurely:

> The first two classes ran off quite smoothly, the lights going off in the middle of the second. However, the third period, which begins at 2:30, saw the pupils rather restless.... Soon it became necessary to lock the windows as the now

> steadily rising wind was pushing them open. Troubled parents called for their children. . . . By three o'clock all means of communication were down. . . . The wind had risen to an unbelievable velocity. . . . Steadily the gale grew stronger and yet stronger. . . . All students had been sent to the auditorium to be away from windows, should they break. . . . A brief lull occurred but when the gale returned it beat with renewed fury at the school. . . . Shortly after five o'clock the wind had abated or so it appeared to those in school. . . . At about 5:30 students who had not been called for earlier begun to leave in groups of two or more. They joined hands and ran against the wind which they discovered was still quite strong. . . .[7]

The Weather Bureau observer in Bridgehampton, Ernest S. Clowes, who was also the author of a full account of the Hurricane's effects on eastern Long Island, provided a special report with detailed observations from noon to 10 P.M. He recorded pressure, wind, and rain conditions at intervals as short as 10 minutes during the height of the storm to two hours toward the end (instead of the expected once-daily observation).[8] His report shows that at noon the barometer was already dropping rapidly and the winds were already blowing at 25 mph. The rain started at 1:15 P.M. and the wind continued to increase steadily up to 90 mph at 4 P.M., followed by a decrease to 60 mph (which was likely interpreted as the lull mentioned in the school account) at 4:30 P.M., and then back up to 90 mph at 4:50 P.M. The roaring at the times of maximum wind must have been tremendous. Winds were 55 mph when power went out at the school around 2:30 P.M., and they were still about 60 mph when the last students were sent out at 5:30 P.M. Compared to the roar of near-100-mph winds, 60 must not have made as much noise and perhaps it seemed like the storm was nearly over, but the winds were still clearly too strong for children to safely walk through.

Spending the night in houses with smashed windows, broken chimneys, and roofs completely or partially missing, or stranded by fallen trees and flooded surroundings, people were in distress, to say the least. Rescue efforts commenced as soon as the weather allowed, and individuals and families were taken to shelters or to the safer homes of family and friends. Survivors also began to show up in downtown areas by their own means. They came in various states of disarray, shivering, clutching each other—some hysterical, some laughing, some desperate to find their loved ones, some injured—all with their own dramatic stories of how they survived. Those who did not suffer any direct loss or who were relatively safe during the storm came out to find a scene of devastation much worse than they expected.

The seawall at Narragansett Pier, southern Rhode Island. In spite of its destruction, it appears that the seawall saved the seaside homes. Photograph by Steve Nicklas, NOS, NGS; courtesy NOAA.

A rowboat travels a flooded city street after the Hurricane of 1938. (Courtesy of the Boston Public Library, Leslie Jones Collection)

Some of the biggest coastal losses were experienced by fishermen, many of whom lost their homes, boats, equipment, loved ones, or even their own lives. Coastal farmers also suffered significant damage, losing their homes, equipment, crops (both harvested and still planted), animals, and in some areas, even the topsoil (and hence productivity in subsequent years). Coastal and rural communities were damaged much worse than cities, where sturdier brick construction was more common. In the cities, however, the amount of debris clogging streets was immense, and many businesses and stores suffered flooded basements and ground levels as well as broken windows at higher levels. Many churches lost their steeples, some hazardously falling through the roof.

As is common with any type of severe flooding in which human structures are involved, the flooded waters teemed with everything from structural debris to various types of animals and a variety of substances. When the waters retreated, many of these problems abated, but standing water of questionable quality was left in ponds and basements, and drinking water reservoirs were equally compromised. City officials instructed people via newspaper, radio, or word of mouth to boil water before drinking or using it to cook. The mayor of Worcester, Massachusetts, warned the public over the WTAG radio station, which was still functioning on an emergency hookup: "Stay off the streets! Watch out for live wires! Boil your drinking water!"[9]

Further inland and well into interior New England, the most widespread damage was the bending, breaking, and uprooting of trees. The trees that were knocked down directly by the hurricane winds caused further damage to other trees and to nearby structures. Inland flooding from rivers and streams was also a big problem in this area. Excess water resulting from the preceding intense rainfall event as well as the Hurricane itself (very little of which could be absorbed into the ground) was draining within the large river basins in any way it could, through streams and tributaries or simply as downhill runoff. All bodies of water within the region, small and large, were consequently experiencing volumes large enough to spill out and flood their surroundings. Some of the first locations to be affected started flooding while the winds were still blowing, which made fighting the flood a difficult task. Once the storm subsided, the frightening and dramatic events were still not over for many towns and cities along the largest rivers, which took a day or two longer to peak, and even longer to go back down to prestorm levels. The Merrimack River, which passes near both Manchester, New Hampshire, and Lowell, Massachusetts, for example, peaked on September 23 and a week later had not fully receded back. The same was true for the Connecticut

River in Hartford. People in these locations, already dealing with tangles of trees and other damage, were frantically racing to secure dams and levees, to build temporary sandbag walls, and to evacuate before the river flooding put them in additional danger. The good news was that many flood-preventing infrastructure improvements had been made after the massive and all too recent spring flood of 1936. Still, flood levels were exceeded at various tributary points, in some cases by 8 to 10 feet above those in 1936. Most of the locations directly on the Connecticut and Merrimack Rivers, however, did not reach 1936 levels. Consequently, many felt thankful, convinced that the river flooding during the Hurricane would have been even worse if not for the recent improvements.[10]

Flooded cities faced an additional danger. Electrical circuits shorted by water can easily start a fire.[11] Once started, such a fire would be very hard to control in the conditions during and after the storm. Massive fires broke out in the business districts of New London, Connecticut, and Peterborough, New Hampshire. New London firefighters faced great challenges in combating the fire:

> Flames burst forth in a business section near the waterfront at about 4:30 P.M. Fire Chief Shipman has explained that water, flooding the building of the Humphrey-Cornell Company . . . short-circuited electric wires. The sparks found ready fuel and the blaze was carried on the wings of a hurricane. All fire-fighting equipment in the city was called out, but firemen had first to hack a path through trees fallen across every street before they could reach the scene. Orders were given to summon fire departments of nearby towns, but all telephone wires were down. . . . By 11:00 P.M. desperate New London firemen prepared to dynamite in an attempt to check the spreading flames. Shortly afterward, the wind veered and swept back over the smoldering area, preventing further loss.[12]

Three hours later, the fire was finally under control. In the end, several buildings, "13 or 14 of the city's largest business and commercial establishments," lay in smoking ruins as the National Guard, Navy, Marines, and Coast Guard patrolled the area.

In Peterborough, a much smaller town than New London, the start of the fire was described as "spontaneous combustion in grain bins" owned by the Farmer's Grain Company. Spontaneous combustion would require an internal reaction within the material to release enough heat to cause ignition. Instead, the most likely cause of these fires (other than the possibility of an open flame directly set to the grain containers) were sparks or discharges occurring between grain dust particles that had become electrically charged by friction as the grains blew around the containers.[13] Regardless of the specific cause, firefighters could not reach the flames because of the floodwaters. Fortunately, the fire did not spread to other parts of town because "buildings were water-soaked by the heavy rains" and embers carried by the wind would not easily rekindle.

New Englanders had never had to respond to this degree of devastation. There was so much to clear, clean, and reconstruct that the burden on towns and communities was truly massive. While some people began to clean up immediately, it took days before the full extent of the catastrophe was even fully realized. Many communities were completely cut off from the rest of the region, with roads and railroads blocked or washed out and telephone and telegraph communication down. Newspapers were equally incapacitated. During one of the biggest stories in generations, in the worst affected areas, there was no power to run the presses and some newspaper facilities had themselves been heavily damaged, mostly by the floodwaters. Reporters, nevertheless, still got out and put themselves in danger in order to do their job. They then typed by the light of candles and lanterns held by colleagues.[14] Within just a few days, newspapers found alternative ways to publish. Just before World War I, the printing process benefited from technological advances that had occurred since the late 1800s.[15] After the Hurricane, some newspapers had to revert to their previous manual processes, some published a mimeographed edition, and some arranged to use the working facilities of other newspapers in nearby less-affected areas. A behind-the-scenes article published by the *Providence Journal*, the largest newspaper in the area most affected by the storm surge, eloquently describes their plight:

> Painfully, laboriously, under unprecedented emergency conditions, the story of the great disaster took shape. The human machine had come through. But the other machines, the mechanical apparatus . . . could not perform their task. The huge presses had been disabled by flood waters, and even if it had been possible to repair them quickly—the job actually took 10 days—they were useless as long as the electric power, their life blood, was dead. Yet, it was imperative to publish a newspaper. It was not merely a question of continuing the newspaper's normal function. . . . With widespread devastation and disruption of all normal public services, the maintenance of some medium of communication had become a vital necessity to the ravaged State. Only thus could military orders, health instructions, casualty lists and news of

catastrophic conditions be made available to the people. Otherwise the story assembled with such tenacity and disregard of danger would be worthless. Hasty arrangements were therefore made to publish the Journal and Bulletin in undamaged newspaper plants outside of Providence. For a week and a half the morning and evening papers were printed in Woonsocket and Boston, respectively. In the midst of the greatest calamity in their history, and of the most acute European crisis since the World War, the people of Rhode Island were supplied with the news.[16]

An Altered Coastline

The remarkable geology and topography of Long Island and southern New England are the combined result of oceanic and glacial forces. The susceptibility of the terrain to these forces, however, can vary significantly. Exposed bedrock (solid rock that forms the earth's crust and is normally covered by the soil) is relatively resistant to erosion. On the opposite extreme, loose sand and gravel are much more vulnerable and easily eroded. It is easy to understand that oceanic tides, currents, and waves can shape the coastline. This can happen via slow and steady repetitive processes over many years or it can be precipitated by sudden forces from events such as nor'easters or the occasional but intense tropical cyclone. Glacial processes can also shape a coastline, but over much longer time scales. During the last period of glaciation of Earth, which came to a halt approximately 10,000 years ago, the ice extended southward over a large portion of the northern half of North America. Over what today is the northeastern United States, the glaciers' southward advance extended to just past the southern New England coast. The slow movement of the ice scraped the terrain, breaking off sediment and rocks. These materials were deposited in an accumulation of unconsolidated glacially scoured soil and rocks known as a moraine. A terminal moraine, left behind after the maximum southward ice extent, usually appears as an elongated hilly ridge. The terminal moraine resulting from the last glaciation can be seen today as a discontinuous line of terrain formed by Long Island, Block Island, Nantucket, Martha's Vineyard, and Cape Cod. Much of the coast of Rhode Island is also made up of the same morainal material and therefore lacks the strength of the bedrock that constitutes most of New England.[17] Roughly stratified sand, gravel, cobbles, and boulders resulting from both glacial deposition and erosional processes (and in some spots from river-mouth deposition) can all be found throughout the coastal area.

The combination of the Hurricane's storm surge, both as the water rushed inland and then when it ebbed back toward the ocean, and the accompany-

TABLE 8.1. Geological Effects of the 1938 Hurricane on the New England and Long Island Coasts

Formation of inlets eroded through bar beaches
Erosion of ocean side of bar beaches
Widening of bar beaches by deposition of material on their bayward side
Erosion of the ocean-facing dunes
Erosion/recession of beach lands and cliffs
Deposition of beach sediments in bays and marshes
Conversion of freshwater ponds into saltwater bays

SOURCE: Nichols and Marston (1939) report in the *Bulletin of the Geological Society of America.*

ing battering waves resulted in an array of effects on the coastal area (see Table 8.1). In only a few hours, this one storm caused more coastal changes than more than one hundred years of natural processes that had occurred since the catastrophic Great Gale of 1815 (see Chapter 9).

In the words of a 1939 *American Geographical Society Review* article, the southern New England outer shoreline presented an interesting combination of "rocky cliffs and, between granite headlands, long stretches of sandy bay bar and barrier beaches surrounded by stable sand dunes knit with grasses and partially covered by low vegetation growth."[18] The low-lying sandy bar beaches that lined the coast were heavily favored as swimming locations as well as for beach-house construction. They were also, of course, extremely vulnerable to stormy conditions, even during "lesser" winter ocean storms. The large barrier beach dunes facing the ocean had had more than a century to build up by the combined effect of wind and waves.[19] They were as large as 30 feet high and 100 feet wide, steep on the ocean side, and gently sloped on the opposite side.

The bar beaches were breached in many places by the Great Hurricane, creating countless inlets that were anywhere from tens to hundreds or even more than a thousand feet wide. Some of these inlets started healing almost immediately by sand deposition and most of them eventually closed up. A small handful, however, remained open a while longer, the most famous being the Shinnecock Inlet, the easternmost inlet connecting the various southern Long Island bays to the Atlantic Ocean through the narrow offshore barrier islands. The maintenance of this inlet has been a source of controversy, as its presence blocks the natural flow of sand along the beach, causing severe beach erosion to its west; but at the same time it provides a very convenient shortcut to the ocean for boaters. Another interesting

Bay Ridge, Rhode Island, homes undercut by erosion caused by New England Hurricane of 1938. Courtesy NOAA.

example is Sandy Point Island, which was not an island prior to September 1938. The bar, which used to connect to Napatree Point (prominently featured in Scotti's *Sudden Sea*), suffered two openings through its narrowest sections, but only one of them closed again. Without fully surveying the difference in geological features and obtaining historical town information, it is hard to know definitively how much of the resulting coastal state was natural and how much resulted from human work, as most towns did put efforts into either filling the inlets or dredging them to keep them open. They also reconstructed seawalls, breakwaters, and jetties, all in an attempt to restore their coast to prestorm conditions. However, the same natural forces that kept the coast in a state of equilibrium during the decades preceding the storm would have continued working to restore the familiar coastal topography. That is, except in cases where the Hurricane's destructive forces were great enough to create a new equilibrium, with entirely new topography bearing no resemblance to what existed before.

Small dunes were completely washed away and large dunes were severely eroded. The water cut channels through them, and the removed sand deposited behind the breaks in fans that stretched hundreds of feet toward the inland side. A large amount of sand was also most likely washed out to sea. The presence of this kind of extreme sediment deposition caused by such major and infrequent hurricanes in the area can be used to determine such occurrences through the last few thousand years, as will be discussed in Chapter 9. The storm surge and waves moved an enormous amount of sand away from the lower beaches (which resulted in considerable erosion on the ocean side of the beaches) and toward the upper beaches (extending them toward the inland or bayward sides). Freshwater and saltwater ponds, which pepper the coast, were made shallower by this additional sand. Wave action, rain, and stream flow, however, washed much of this sand back toward the lower beaches after the storm. Natural dune rebuilding also commenced immediately, but it would take many years or even decades for the largest ones to reach sizes comparable to their prestorm state. Ponds and swamps near the coast were overwhelmed by the ocean water. The freshwater ponds thus temporarily became saltwater ponds or bays, but they also quickly started to revert to their naturally fresher state.

The water didn't just move remnants of homes and sand. Large boulders were carried quite a distance inland on the low-lying shores, well away from their original locations, "beach cobbles were thrown over a 9-foot scarp and covered a lawn a hundred yards inland, and fine shingle was carried an equal distance up and over a 20-foot cliff by wind and spray."[20]

An effect less likely to heal and revert back to prestorm conditions was the severe topsoil erosion suffered throughout the coast. The higher sea level and large waves were able to reach less resistant materials normally affected only by gentler and much slower weathering processes. These higher elevations were therefore eroded more severely by the storm than their lower counterparts. The scouring and battering not only removed all the covering vegetation in some places but also caused the coastline to recede by tens of feet in various locations. Coastal cliffs consisting of exposed bedrock were only slightly eroded, but those made of glacial deposits were severely affected. Near Watch Hill Point in Rhode Island, for example, a 50-foot cliff composed mainly of stratified sand was stripped of its vegetation and eroded as much as 30 feet inland even though it was protected on the ocean side by a concrete seawall.[21] The topsoil loss, however, was severe even in more resistant areas. In places like Newport, Rhode Island, for example, the soil covering was completely removed down to the bedrock for several feet inland.

A year after the Hurricane, it was believed that soon enough, within a few years at the most, the beaches and the coast in general would be essentially in the same condition as they were before. For the most part, this was the case. New vegetation grew and sand deposition closed the storm-carved channels breaching the bar beaches and, with the help of the wind, started rebuilding the dunes. Nevertheless, some permanent scars did remain.

Farms and Forests Destroyed

The high storm surge and the substantial water runoff resulted in extreme soil erosion in the immediate coast and in many open fields throughout the region. Agricultural fields that had been harvested recently and had no vegetation cover were especially affected, the anchoring plants having been scraped away together with the soil in some areas. Additionally, the large amounts of salt either brought in by the storm surge or simply blown inland with the wind (as attested by those who reported they could taste the salt in the rain) took care of damaging much of the surviving vegetation. For miles inland from the ocean, the little plant life remaining intact looked like it had been exposed to a killing frost.[22] Much of the vegetation submerged in the salty water (grass, crops, gardens, wild bushes, or trees) died soon afterward. Farther inland, the effects of the salt were much less noxious to plant life and much less noticeable, although a salty residue was observed as far inland as a hundred miles or even farther away from the ocean in central New Hampshire and Vermont.

Crops of all kinds were ruined. Anything that was still planted or stored was damaged, knocked down, flooded, or sprayed by salty rain. Many potato farms near the ocean were either completely washed out, covered by sand, or so salt-ridden that the potatoes would rot soon after being dug out. Harvested tobacco kept in barns was a total loss.

Farther north, among all the tree loss, entire groves of red and sugar maples were downed, severely affecting the maple syrup production for years after the storm. One of the most ubiquitous losses was the apple crop. Entire orchards were downed, not to mention the millions of apples knocked down by the wind. While some of the fruit were damaged, a surplus survived well enough for consumption. Orchard owners not only saw their trees lost but also faced crashing prices due to this excessive supply. An account of the storm by the Federal Writers' Project (a program of the Works Progress Administration [WPA], described below) explains what was done to alleviate this:

> A call went through New England to save the growers from ruin. The people responded. Restaurants are serving their costumers cider instead of water; a variety of apple dishes is being featured; housewives have been buying more apples for the health of their families and the preservation of New England orchards; the government has bought hundreds of carloads for the needy.[23]

Woman busy preserving food after the Hurricane of 1938. (Courtesy of the Boston Public Library, Leslie Jones Collection)

The heavy tree loss within the White Mountain National Forest immediately caught the attention of the U.S. Forest Service. It became clear right away, however, that it was not just the protected areas but the entire region that was going to need serious restoration work. The agency quickly intervened with two very different but related goals: fire hazard reduction and timber salvage. The first was considered more urgent and commenced immediately. The second was unprecedented and, while time sensitive—as the wood needed to be collected before it became damaged— took longer (about a month) to organize and begin execution.

There was so much brush on the ground and so many trees down that it was extremely hard to contain any naturally occurring wildfires. Five years after the Hurricane, the Forest Service published a very detailed report of all their operations following the storm; this is how it described the fire hazard:

> From a region of moderate fire hazard with comparatively favorable conditions for fire control, the area became a patchwork of blown down forests which, being mostly pine, represented a new high in forest fire hazard. The massed pine tops were highly inflammable, and the tangle of fallen trees

> would have resisted any normal fire control efforts. This critical condition presented a problem of such magnitude that its diminution was considered more urgent than the problem of salvage. It not only threatened further loss of log values involved in the blown down material, but it constituted a major threat to remaining property and to human life.

The biggest concern was that an extended period of dry weather (such as the one that happened after the Hurricane) followed by high winds would translate into any naturally occurring small fires easily propagating throughout the downed timber and quickly becoming uncontrollable. To make matters worse, much of the infrastructure formerly in place for fire control was gone. Watchtowers where Forest Service personnel or local foresters once stood watch to detect any nascent fires went down with the trees. Additionally, many trails and roads were blocked by trees, making firefighting equipment access impossible. As many as 2,000 or more fires were expected within the ravaged area during the next year, 1939.[24] The majority would most likely occur within the spring fire season, beginning after the snow melted, ending when the vegetation was completely grown, and peaking from late March to May. It was important to act as quickly as possible, before winter weather made the cleanup work much more difficult. Many fires did indeed occur, although there is no detailed record of how many.

The intensity of the fire hazard varied from slight to extreme from one location to the next. The local damage was surveyed and the danger evaluated individually for each town, matching the already established boundaries of firefighting responsibilities when possible. In some areas, conditions were so extreme that they far exceeded the experience of any of the local firefighters, making it necessary and desirable for an agency with many more resources to step in. The Forest Service oversaw federal workers and local volunteers as they replaced fire towers; opened roads, fire lanes, and trails; and cleared downed material from roadsides, villages, dwellings, and mills.[25] The clearing was done throughout private, town, state, and federal lands alike, but when private property was involved, written permission was required to avoid liability. Total clearing and cleanup, however, was virtually impossible and could never be the realistic goal; it was only done for a minimum number of high-hazard locations where human life or valuable property was imminently threatened. Materials not slated for salvage were put in large brush piles and burned under controlled conditions. Countless small fires did break out during the cleanup operation, and had to be constantly put down by whomever saw it first or had the best access. Oftentimes road

The pavilion at Easton's Beach, Rhode Island, being burned intentionally prior to beginning cleanup and rebuilding. Photograph by Steve Nicklas, NOS, NGS; courtesy NOAA.

construction, power, or telephone crews or local volunteers performed the initial firefighting, with town and federal crews then taking over. For fires of larger proportions or when special equipment was needed, the appropriate state forester office would provide help and out-of-state resources were additionally obtained when necessary.

The Forest Service had never in its history been involved in timber salvage operations—and one of this magnitude had never been attempted before by anyone—but it seemed the agency was in the best position to take care of the massive problem most efficiently "in order to prevent the development of serious losses through insect damage, fungus diseases or exposure."[26] The New England Timber Salvage Administration (NETSA) was therefore quickly created under the umbrella of the Forest Service specifically to manage all aspects related to this complicated effort. The new, specialized agency was in charge of all timber salvage and related activities, including administrative and financial management.

The amount of downed timber was equivalent to what would normally be logged in 5 to 10 years, producing an excessive supply that would have certainly driven market prices low enough to cause financial losses or at best negligible profit to landowners. "It became increasingly apparent that the federal government should participate in some manner by making financial assistance . . . to allow local people to realize a maximum return from what

THE WOMEN OF TURKEY POND

Four years after the Hurricane, the Northeastern Timber Salvage Administration (NETSA), trying to meet its 1943 deadline to process the piles of timber that had accumulated everywhere and facing a shortage of male labor due to the war, established a sawmill at Turkey Pond in Concord, New Hampshire, to be operated by women. Turkey Pond, close to many heavy blowdowns, was the receptacle of 12 million board feet of white pine. This was the largest deposit of salvaged logs anywhere at the time. The logs floated like matchsticks almost as far as the eye could see. By the time the sawmill was built (with careful attention to safety measures), most of the salvage effort had ended for the rest of New England. An internal publication of the Forest Service described how well the "experiment" was going:

> The "female" mill at Turkey Pond is going along nicely. It is most surprising and gratifying to see the way those gals take hold of the job. In addition to the jobs we anticipated women could handle we have found them capable of rolling logs on the deck, running the edger, and . . . even, running the head saw. Maybe it will be possible to actually man a mill 100 percent with women sometime in the future.

New Hampshire historical marker number 184, which can be found at Turkey Pond, adds: "Led by sawfiler Laura Willey, the women proved them-

could have been a nearly total economic loss." The federal government thus purchased salvaged material at what it considered reasonable prices with the intention of selling on the market as demand arose. Even though some of it was indeed sold in this way, the stockpile of government-owned lumber was ultimately extensively used during World War II.

Land owners brought their salvaged logs to designated locations via horse- and oxen-drawn carts and tractors. The wood was then measured and either dumped in a pond (more fragile softwoods, mostly white pine and some spruce) or piled on fields (hardwoods and hemlock). A total of 260 ponds and 675 fields were used as storage sites with the intention of fully processing them in a few years.[27] Salvage operations continued until the summer of 1941, and all administrative operations of NETSA were ceased by 1943.[28]

Female lumberjacks carrying logs at Turkey Pond, New Hampshire, as part of an experimental project to saw up 7 million feet of lumber from the 1938 Hurricane. Courtesy National Archives and Records Administration, Boston.

selves to be an exemplary crew. 'Snow, rain, or sub-zero weather never slowed them up,' wrote one Forest Service manager." Their story has been told in a 2010 book titled *They Sawed Up a Storm* by Sarah Shea Smith.

Leisure activities that New Englanders and visitors were used to enjoying in the forest and mountains were obviously affected greatly. With blocked roads and trails and the high fire danger, many state parks and the White Mountain National Forest were at first closed to the public. The closures, however, were not long lasting and did not extend across the region. The upcoming hunting season, surprisingly, was not cancelled and some did venture out into the tangled woods.[29] Many locals and visitors from the southern portions of the region ventured into the woods out of a seemingly morbid curiosity to see the damage. Indeed, tourism was important to the economic recovery of the region, and there were internal Forest Service talks concerning the quieting of rumored fears that one spark would "roar through New England faster than the hurricane itself," as a popular magazine allegedly put it. Although the fire danger was still very real, the goal was to not scare tourists away from New England the following summer; but once the tourists did come, the agency hoped to "scare the hell out of [them] so that [they would] be careful with fire in the woods."[30]

Interestingly, the massive removal of downed trees, deemed necessary for protecting life and property, might have itself severely interfered with the natural recovery of the forest. Dramatic changes in forest composition followed the 1938 Hurricane, and there were also major hydrological changes. A significant increase in discharge was observed within the affected watersheds, such as those of the Connecticut and Merrimack Rivers.[31,32] These two effects have long been considered natural consequences of the catastrophic forest blow-down, but ecological studies performed 50 years later suggest otherwise.[33] The Harvard Forest Long Term Ecological Research program performed an experiment that simulated the type and amount of damage done by the Hurricane and then measured various aspects (tree population dynamics, vegetation structure and composition, individual organism growth and productivity, and even chemical fluxes above and below ground) of the forest recovery for many years afterward. Many of the downed trees actually survived, sprouting leaves the following growing season. Additionally, there was no significant change in the species composition of the forest, the soil chemistry, or the moisture levels. While the structural damage to the forest was enormous, most other forest mechanisms remained intact. The downed material that did not survive provided shelter to wildlife and eventually provided renewed nutrients to the soil, maintaining net chemical and moisture budgets in the process. This was only a limited-area experiment, but it demonstrated that it is very possible many of the large forest changes beyond the initial Hurricane damage were caused by the massive removal of downed material rather than the downing itself.

Red, White, and Blue Relief

The Forest Service certainly had the biggest role in directly dealing with the trees, but it was not the only or even the main agency or organization, federal or otherwise, involved in dealing with the aftermath. Today, a disaster like the one caused by the Hurricane would be managed in great part by the Federal Emergency Management Agency (FEMA), which for the last few decades has been in charge of responding within the United States and its territories to disasters that are of a scope large enough to overwhelm local and state resources. No such agency existed in 1938. The involvement of the federal government in the relief efforts following the 1938 Hurricane represented the first time anything of its kind occurred on such a grand scale. It was not, however, the first time that the government provided any type of disaster relief. Starting in the 19th century, U.S. government involvement in natural disaster relief occurred on a case-by-case basis via Congress-approved measures. The awarded relief mostly consisted of tax-deadline extensions or remissions for those affected by the specific disaster. During the Great Depression, the government also started to approve some financial relief measures such as loans or road reconstruction, which previously was the sole responsibility of individual states or towns.[34]

The majority of the cleanup and reconstruction job following the Hurricane was performed by an agency that no longer exists, the Works Progress Administration (WPA), which was the largest and most ambitious of President Roosevelt's New Deal agencies created during the Great Depression. The WPA employed millions of workers, mainly to carry out a great deal of infrastructure improvements—to roads, dams, public parks—and any other type of public work that needed attention. It was also involved in literacy, drama, and art projects, encouraging unemployed writers, actors, and artists to practice their craft in a very weak economy where they would have otherwise struggled. All WPA projects and resources were put to use in September 1938.

When the Hurricane came, there was no preexisting emergency relief plan of any kind. One would expect this would have interfered with a prompt and appropriate response, but that was certainly not the case. The WPA already had local men in its rosters all across the region, and it had no problem recruiting additional men from the ranks of the plentiful unemployed. It took practically no time to call troops into action—the weather had barely cleared when WPA workers were already on the job. One of the first duties within the worst-affected areas was the recovery of bodies. Parties of searchers picked "their way among the shattered timber, lifting boards and pushing aside debris in the hope of discovering the bodies of those who perished."[35] Workers also cleared roads of trees and debris, repaired washed-out roads and bridges, took down unsafe structures, and performed any needed service, from helping the Red Cross with shelters to manufacturing clothes for those who needed something to wear other than the clothes on their backs. Less physical work, such as making clothes, provided the perfect opportunity for women and disabled men to contribute, and their involvement was proudly highlighted, even if described in a way that would be considered questionable today: "Crippled men and women joined the WPA sewing projects in making thousands of garments for flood victims."[36] WPA workers constructed

FEMA

The Federal Emergency Management Agency (FEMA) was established in 1979 by executive order of the president of the United States (then Jimmy Carter), in the process absorbing a series of existing federal agencies with a variety of relief responsibilities under one independent agency. In 2002 FEMA was put under the new Department of Homeland Security, created after the September 11, 2001, attacks with the overarching goal of protecting the United States from within its borders (by fulfilling its missions to prevent terrorism, enhance security, and ensure resilience to natural disasters among other emergencies).

sandbag walls where the rivers continued rising and assisted the Forest Service cleanup groups where trees needed to be cleared and salvaged. The Federal Writers' Project, also part of the WPA, documented both the aftermath of the storm and the relief activities through a geographically organized and somewhat poetic pictorial account.[37] Thanks to their account we know that in Rhode Island "day after day, WPA and volunteer workers and National Guardsmen pushed on doggedly with their task of freeing the land of the weight of smashed buildings and wreckage dumped by risen waters," and in Lowell, Massachusetts, "windowless factories gaped down on exhausted volunteer workers." The WPA also set up shelter playrooms to entertain the hundreds of displaced children, and the Federal Music Project (also part of the WPA) provided musical entertainment.

While the WPA took care of the many aspects of public reconstruction and services, the humanitarian side was in the hands of the Red Cross, which had a massive role in the post-Hurricane relief. The American National Red Cross is a volunteer-led, donation-supported humanitarian organization founded in 1881 (as an indirect result of the American Civil War and inspired by the international organization of the same name that was created in Europe in 1863). It is not a government agency but, after providing relief for various disasters during the late 19th century, its authority to do so was formalized in 1905 by a congressional charter giving it the "obligation to continue and carry on a system of National and international relief in time of peace and to apply the same in mitigating the sufferings caused by pestilence,

WPA workers and rescue squads search for bodies and survivors at Shawomet Beach, south of Providence, Rhode Island. Photograph by Steve Nicklas, NOS, NGS; courtesy NOAA.

famine, fire, floods and other National calamities, and to devise and carry on measures for preventing the same."[38] It has since been involved in countless disasters and emergencies, big and small, providing immediate relief and a great variety of humanitarian aid services to those affected.

Red Cross activities following the Hurricane covered anything and everything related to relieving human suffering. Chapter members, for instance, even helped in the continuing rescue of those trapped or marooned and evacuating those in endangered areas or unsafe structures (activities that are no longer part of the Red Cross responsibilities as dictated by the United States National Response Framework, which currently provides guiding response principles for domestic disasters and emergencies, including the roles and responsibilities of each involved organization).[39] Red Cross nurses applied first aid and inoculations, and they assisted with medical care at shelters and hospitals. An army of volunteers organized by the agency helped with feeding and clothing those who were unable to return to their homes. The Red Cross also served as a citizen information network. Requests for information on the fate and whereabouts of loved ones were made in person or via telegraph (where possible) or amateur radio and any news that could be verified, good or bad, was relayed back. Inquiries came not just from the Northeast but also from throughout the nation.

To accomplish these tasks more effectively, Red Cross Hurricane relief headquarters were set up in Providence and an entire organizational infrastructure was put in place. The Red Cross had liaisons and representatives at the state and community levels facilitating communication and coordination in all directions, serving as the pipeline that referred those in need to appropriate federal, state, or local services or resources. The agency additionally provided its own support. A workforce of case workers and their administrative staff were brought in to assist in the massive effort.

Case officers, working closely with each affected family, determined how much was lost and what it would take for the family to recover. This did not mean replacement of all their losses, but the minimum necessary for them to get back on their feet. The goal was to fully utilize the family's own resources (both of labor and monetary savings) first and then use their borrowing power to take advantage of the various different types of loans available (to those who had sufficient equity or showed enough potential for repayment). For those whose combined resources were insufficient, the Red Cross covered the gap with grant awards. These were not loans but gifts with no repayment expected. A custom plan was then created for each individual family that could include anything from rebuilding the home to replacing tools necessary to earn a living to providing rehabilitating medical treatment or prostheses to retraining the breadwinner in a different trade.

The emergency relief program, including shelters (which held, fed, and generally took care of 15,000 people) and medical attention (first aid for 6,000 patients and inoculations for 1,000), was completed by October 15. The full disaster relief operation, however, lasted for three months, during which a total of nearly 13,500 families were serviced by individual plans. Any administrative work remaining by the end of January was moved from Providence to Washington, D.C. (to the Red Cross headquarters), and a skeleton staff was left behind to clear any leftover relief work, which mostly involved finalizing construction that had halted during the harsh winter months. The total cost of the operation (see Table 8.2), funded in part by National Red Cross reserve funds and in part by donations received specifically for this disaster, was approximately $1.7 million (which would correspond to approximately $27.7 million in 2012 dollars).[40]

Even though the Red Cross, WPA, and the Forest Service performed the bulk of the heavy lifting, to only describe their role would be to severely misrepresent the combined efforts that occurred everywhere in the region. Individuals, cities, towns, state agencies, private companies, and the federal government were collectively involved in rescue, recovery, cleanup, and reconstruction. There was a sense of responsibility and accountability in the air. Everyone who came out of the ordeal unscathed went out to help those who did not, and even those who suffered severe losses helped others in need. Strangers who would never have known each other's names, as well as friends and loved ones, shared dramatic stories of survival. Scenes like these might be less common today, as more thorough evacuations and personal safety concerns might result in far fewer survivors being present to interact and help one another in severely impacted areas.

Each community's authorities went into action as soon as there was sign of trouble. Agencies helped the public in myriad ways, sometimes blurring the lines of their responsibilities as typically defined. Police and fire department personnel were busy rescuing those in immediate danger. Local and state police as well as National Guardsmen patrolled streets to discourage looters and to provide any needed help on the spot. Firemen used their pumps to clear inundated basements and cellars and their flood lights to allow rescue, cleanup, and construction work to continue through the night. The Coast Guard and the National Guard quickly joined in, providing Army cots and tents for relief workers. The Coast Guard rescued stranded people even inland. Telephone and telegraph operators worked full-time keeping

TABLE 8.2. Red Cross Services and Expenditures

Contributions	
American National Red Cross	$751,251.75
From Organizations and Individuals	
Continental U.S.	$928,034.34
Insular (Panama Canal Zone) and Hawaii	$536.65
Foreign	$2,177.26
Total	$1,682,000.00

Services and Expenditures

Aid Type	Families Requesting	Families Receiving	Total Cost
Rescue, Transportation, and Mass Shelter[a]		3,332	$21,545.62
Food, Clothing, and Other Maintenance	6,780	5,638	$110,223.08
Building and Repair	8,804	4,871	$831,192.61
Household Furnishings	4,949	3,181	$201,374.20
Medical, Nursing, and Sanitation[a]		471	$23,752.42
Farm Supplies, Livestock, and Equipment	1,402	356	$19,573.37
Occupational Training, Equipment, and Supplies	1,773	812	$145,433.67
Family Service[b]			$221,363.89
Administration and Accounting[c]			$95,636.11
Amount set aside for unknown and uncertain awards[c]			$11,905.03
Total			$1,682,000.00

SOURCE: New York–New England Hurricane and Floods 1938, The American National Red Cross, Washington, D.C.

[a] Not a service that was specifically requested, only received.

[b] No data on families that requested or received this service exist.

[c] Not a service that was requested or received.

relief crews in communication (where these services were still available). An entire range of altruistic organizations came out to help, from the Salvation Army and the National Legion to church groups and the Boy Scouts of America. Institutions and their staff, business and store owners and their employees, and regional and national companies rushed to make the repairs necessary to resume needed operations. Railroad, power, telegraph, and telephone companies were especially affected, and their repairs were essential to the relief effort itself and to the returning of normalcy to the region. One can find various reports and special publications and newsletter editions of some of the many local companies that existed at the time, such as the New England Telephone and Telegraph Company, Southern New England Telephone Company, Cape Telephone Systems, and the New England Power Association. It was clear from these documents repair crews were called in immediately from near and far, "from as far south as Florida and as far west as Nebraska," "from Georgia to Michigan," and "from as far as the Dakotas."[41] Additionally there were reporters documenting the events, scientists and amateur observers measuring meteorological and hydrological conditions, foresters and volunteers surveying damage. In summary, the post-Hurricane relief effort was truly extraordinary.

In the face of disaster, the people of the region endured shock and sadness and uncertainty about what the future would hold, but also met the challenge with hope, courage, pride, and diligence to do what needed to be done. The resilient New England spirit survived. Boys climbed the slanted trees, teenagers took turns towing each other to water ski in shallow flooded areas, and adults gathered for companionship, comparing tales and even eventually finding some reasons to laugh. Somewhat humorous stories surfaced during the next few months, including, for example, the gentleman trying on hats that floated past him outside a store in downtown Providence. A recurring theme in reports, newspaper stories, witness accounts, and personal letters is that of "having much for which to be thankful." New England and Long Island were looking "forward, not backward."[42]

9

PAST, PRESENT, AND FUTURE

Previous New England hurricanes, the next big one, and the 1938 legacy

The meteorological life cycle and immediate aftermath of the Great New England Hurricane do not tell the entire story. Natural disasters of this scale have complex and long-lasting impacts, and our account is not yet complete. The long-term legacy, remaining scars, and lessons learned deserve discussion, as does the possibility of such a disaster happening again (or whether it already has).

Those living on Long Island and in New England in 1938 had never experienced a storm as powerful as the one that arrived on September 21. Even as the wind strengthened and the waters rose, it took time for most to realize the seriousness of the storm. The last hurricane to hit northeastern shores had landed during the previous century. Further, Florida had just braced for an impact that never materialized, but Florida could be expected to get hurricanes, and the idea that the same storm could endanger states far away to the north was not a consideration. Today, we know better. After witnessing storms like Carol (1954), Gloria (1985), Bob (1991), and Irene (2011), combined with increased public awareness of weather phenomena and in-depth coverage of those events by the media, we know that a New England landfall not only can but will happen again.

A Brief History of Intense New England Hurricanes

How Often Do They Happen?

The likelihood of direct hurricane landfall somewhere on the coast of the northeastern United States in a given year is not high when compared with the Gulf of Mexico, Florida, and the Carolinas; the likelihood of the region feeling the effects of a major hurricane is expectedly smaller. Focusing specifically on the Long Island and New England coasts, the odds are even smaller (see Tables 9.1 to 9.3, which contain regional frequency and probability estimates) and almost negligible for a category 3 or higher storm. To the best of our knowledge, no portion of the region has ever suffered the effects of a category 4 or 5 hurricane, and storms of category 3 strength have made landfall only a handful of times and only on the southernmost coastlines of the region.

During the 20th century, there were only three category 3 landfalls in New England (one in 1938 and two in 1954). There are various reasons for this lesser frequency. The environment in which the storms move tends to become less favorable (see Chapter 4) for allowing and maintaining a high-intensity hurricane at these higher latitudes. This is also when and where, if the right meteorological conditions exist, many of these tropical systems undergo an extratropical transition (see Chapter 7), resulting in larger but normally weaker systems. Additionally, even though the atmospheric currents in which the storms move can vary significantly from case to case, they often result in a track that recurves over the North Atlantic, away from the northeastern United States and eastern Canadian regions (see Chapter 4).

The National Hurricane Center (NHC) maintains a Tropical Cyclone Climatology website (www.nhc.noaa.gov/climo) containing maps and charts depicting tropical cyclone distribution, frequency, seasonality, and other statistics, all in one way or another based on the data contained within HURDAT/2. One of the products shows estimated return periods (the length of time between occurrences) for a hurricane of any strength, as well as for major hurricanes (categories 3–5). The values are determined by a statistical model named HURISK (Hurricane RISK),[1] which uses annually updated data to calculate, among other statistics, return periods for storms passing within 50 nautical miles of each location. According to the model results, Long Island and southern New England average decades or even longer between intense hurricane strikes. The recurrence, nevertheless, can be highly irregular: there can be years with more than one significant tropical system affecting the area, and long periods of time without an especially noteworthy

TABLE 9.1. Region-wide Statistics from Simple Averages

New England/ Long Island Storms	Total	Yearly Chance[a]	Recurrence[b] (years)	Largest Period[c,d,e] (years)
Most devastating[f]	1	0.6%	162	-
Major hurricane[g]	7	4.3%	23	69
Hurricane	19	11.7%	8–9	28
All[h]	58	35.8%	2–3	10

NOTE: Calculated by simple counts of storms affecting New York and New England (CT, RI, MA, NH, and ME) appearing in HURDAT records from 1851 to 2011. The equivalent updated information is not included in HURDAT2, but in a separate file, a Chronological List of All Continental United States Hurricanes, which identifies the states affected by each storm and the Saffir–Simpson category experienced in each state.

[a] Total number of storms divided by total of 162 years.

[b] Number of years divided by the number of storms.

[c] Maximum number of years in between storm occurrences. Minimum number of years in between occurrences is zero for all cases, since for all categories (except most devastating) there have been at least two in at least one year.

[d] Only one storm in the "most devastating" category exists in the HURDAT/2 records, the Great New England Hurricane of 1938; therefore, maximum and minimum time in between occurrences is not computed.

[e] 1869–1938 for major hurricanes, 1896–1924 for all hurricanes, 1944–1954 for all storms.

[f] Major intensity (category 3) and region-wide devastation.

[g] Category 3 or higher, although no storms of category 4 or 5 have ever affected the region.

[h] All storms of tropical origin making landfall over the region, including hurricanes, tropical storms, tropical depressions, and storms that have undergone extratropical transition.

storm. Just as in every hurricane season there are periods when conditions are more or less favorable for tropical cyclone formation, there are also years or even entire decades with more or less favorable conditions. This means that statistical analyses dealing with frequency over a period of time, though very informative, do not tell the entire story about the chances of a hurricane striking during any given hurricane season.

The Tropical Meteorology Project at Colorado State University (CSU) produces seasonal hurricane forecasts that have been the subject of ample media attention for many years. Every year before and during the hurricane season (in December, April, June, and August), the group publishes its predictions, updated based on environmental conditions determined to correlate well with the seasonal hurricane activity. Their *forecast* terminology should not be confused with the day-to-day predictions that NHC forecasters generate of active tropical cyclones. The predictions are more in line with outlooks of potential weather occurrences over long periods of

TABLE 9.2. Regional Statistics from NHC Climatology

Region[a]	Return Period[b] (years)		Strike Density[c] (total/111yrs)	
	Hurricane	Major Hurricane	Hurricane	Major Hurricane
Northern New England	25–30	121–290 53–120 (NH)	0–2	0
Southern New England and Long Island	17.24	53–120 33–52 (LI and RI)	5–6	2–3
Cape Cod	12–16	53–120	7–9	2–3
Southern Florida[d]	5–7	14–22	20–32	12–15

SOURCE: From return period and strike density maps in the NHC's Tropical Cyclone Climatology page (http://www.nhc.noaa.gov/climo/).

[a] Regions subjectively chosen by examining return period (range of number of years in between occurrences) maps.

[b] Return period for hurricanes passing within 50 nautical miles of coastal locations in the specified region, calculated using 1987 HURISK model with data up to 2010.

[c] 1900–2010 total hurricane strikes per county from NOAA Technical Memorandum NWS NHC 46.

[d] Southern Florida, the region with the highest hurricane frequency in the United States, is included for contrast purposes.

time (months, seasons) that are beyond predictability by normal numerical weather forecasting models. These outlooks do not directly translate into a seasonal landfall probability. The CSU group, however, also includes an overall index of expected activity for the year, the Net Tropical Cyclone Activity (NTCA), as part of its seasonal statistics. The NTCA, as its name implies, gives a general idea of how active a particular season is expected to be. As a separate project, the group also calculates probabilities of landfall by any named storm (tropical storm or stronger), hurricane, and major hurricane for various regions (including southern and northern New England as separate regions), as well as individually for each coastal or near coastal county. It refers to these general probabilities as the "climatology" for a specific region or county (numbers for New England are shown in Table 9.3), but by using the NTCA as an adjusting factor, it also calculates annual probabilities. This is an indirect way to account for how favorable or unfavorable environmental conditions for hurricane formation are for a specific season.

Based on the adjusted climatological probabilities, the likelihood of a hurricane making landfall anywhere in New England or Long Island on any given year is small, though not insignificant (except for the case of a major

TABLE 9.3. Regional and Select County/City Average Probabilities from CSU Project

	Named Storm[a] (% chance)		Hurricane (% chance)		Major Hurricane[b] (% chance)	
	Yearly[c]	50-yr[d]	Yearly	50-yr	Yearly	50-yr
Northern New England[e]	5.6	94.9	3.2	80.4	0	<0.1
Down East ME[f]	1.3	49.5	0.8	31.5	<0.1	<0.1
Portland	0.5	22.8	0.3	12.4	<0.1	<0.1
NH Seacoast[g]	0.3	14.4	0.2	8.3	<0.1	<0.1
Boston	0.1	5.6	0.1	3.2	<0.1	<0.1
Southern New England and Long Island[h]	15.4	>99.9	9.2	99.4	4.4	89.9
Cape Cod	2.7	75.1	1.6	54.9	0.7	30.9
Eastern Long Island	5.6	94.6	3.2	81.1	1.5	53.7
Bronx, NYC	0.3	12.4	0.2	7.3	0.1	3.5
Southern Florida[i]	36.7	>99.9	27.9	>99.9	13.9	>99.9
Miami	8.1	98.8	5.8	95.5	2.7	75.3
FL Keys[j]	22.3	>99.9	16.5	>99.9	7.9	98.6

SOURCE: From CSU's Landfalling Hurricane Probability Project. Probabilities are calculated by using a Poisson regression model. The numbers shown represent the average or climatological probabilities (not adjusted for any specific seasonal activity).

[a] Including tropical storms and hurricanes of any intensity.

[b] Category 3 or higher.

[c] Probability of occurrence on any given year.

[d] Probability of occurrence within any 50-year period.

[e] Region 11 in the CSU study, including coastal counties from approximately Boston to easternmost Maine.

[f] Easternmost portion of coastal Maine (Washington County).

[g] Small coastal portion of New Hampshire (Rockingham County).

[h] Region 10 in the CSU study, including coastal counties from NYC to Cape Cod.

[i] Region 6 in the CSU study is the one with the highest probabilities. Included for contrast.

[j] Monroe County, within region 6, is the county with the highest probabilities in the entire United States.

hurricane in northern New England, which is very small). In contrast, the probabilities for southern Florida, including Miami and the Florida Keys—with highest chances anywhere in the nation—are dramatically higher.

The contrast between seasons can be clearly depicted by two extreme examples. On one hand, the 1954 season brought two major hurricanes, Carol and Edna, to the Northeast. This was the infamous hurricane season that precipitated the creation of the NHC as its own agency (as briefly described in Chapter 2). On the other hand, the last three decades of the 19th century and the first three of the 20th went by without any category 3 landfalls. The timing of this lack of activity is very significant, because it exactly corresponds to the decades since the creation of the agency in charge of hurricane forecasting, meaning that even the most experienced Weather Bureau forecasters had never seen intense hurricanes in the area. Very often systems of tropical origin came inland or very nearby, perhaps close enough to be felt, but the great majority of them were of lesser intensity, weakened to tropical storm or tropical depression strength, many having transitioned to extratropical. To New Englanders, these storms would not have felt much different or caused damage any more severe than would result from the extratropical cyclones that regularly affect the area throughout the year. The three-plus decades starting in 1900 were especially tame (see Table 9.4 and its accompanying discussion), with only two hurricanes, one in 1924 that sideswiped Cape Cod and one in 1934 that crossed Long Island. Neither of these two systems, however, was nearly significant enough to make forecasters think that anything stronger than a regular nor'easter could affect the region. The Northeast storms that occurred on their watch had either weakened significantly or turned away before they became a problem, as expected.

It turns out that the very year before the founding of the agency as a military division in 1870, two significant hurricanes did come to New England. The first was most likely a category 3 storm and the only major hurricane in the 1800s appearing in the HURDAT/2 records (which start in 1851). The September Gale of 1869, as it is known, was of great intensity, but its impacts were somewhat stunted by its compact size (inferred from the narrow area and short duration of strong wind observations) and its timing (during low tide), counteracting the storm surge.[2] It was, nonetheless, described as the most significant hurricane to affect the region since 1815. Just one month later, a highly anticipated storm known as Saxby's Gale (also known as the Great New England Rainstorm or Down East Hurricane of 1869) affected with unusual force a region normally spared by tropical systems: northern Maine and southeastern Canada.[3]

It is therefore easy to understand that the arrival of a hurricane, especially an intense and widely devastating one affecting the entire region, would not have been considered a likely event in September 1938. Nevertheless, in sharp contrast to the apparent lack of awareness of past storms, lists of previous New England hurricanes were included as part of many of the reports and articles published soon after the 1938 storm, almost as if the readers needed

SAXBY'S GALE

In November 1868, a year before what was most likely the equivalent of a category 2 hurricane affected Maine and eastern Canada, Lt. Stephen M. Saxby of the British Royal Navy sent a message to London predicting a storm of "unusual violence, attended by an extraordinary rise in tide" on the morning of October 5, 1869. An amateur astronomer, he based his prediction on the impending alignment of the moon over the equator. The prediction was widely discussed in scientific and social circles with great anticipation, especially in Halifax, Nova Scotia, where a local named Frederick Allison predicted Saxby's Gale would hit. For Halifax, it was a close miss, as a storm indeed hit but with a track that kept it west of the city. The Hurricane caused extensive damages along the Bay of Fundy and the coast of Maine.* Of course, it is not possible to predict a specific storm an entire year in advance. The fact that one actually occurred was a coincidence, albeit an interesting one.

* Ludlum (1963).

to be convinced of the matter. The hurricane report from the Weather Bureau published in the September 1938 *Monthly Weather Review*, for example, stated simply that "many storms of tropical origin have previously affected the New England States" and then went on to name some of them (several of which are described here).

There have been, of course, other notable New England hurricanes besides the three massively devastating ones that include the 1938 Hurricane and those in 1635 and 1815. As mentioned in previous chapters the 1743 Benjamin Franklin's Eclipse Hurricane led Franklin to realize that the direction from which a storm is moving and the direction from which its winds blow are not necessarily the same, and the Norfolk and Long Island Hurricane of 1821 led William Redfield to conclude that hurricanes are rotating "whirlwinds." There were also the two abovementioned 1869 storms, and Ludlum's compilation of early hurricanes contains descriptions of several other storms that affected New England, 10 to 15 of them arising before the HURDAT/2 records start in 1851.[4] Since 1938, additional notable storms have affected the region (Table 9.4): for example, the 1944 Great Atlantic Hurricane, Carol and Edna (both in 1954), Donna (1960), Gloria (1985), and Bob (1991), all of

TABLE 9.4. "Modern" New England and Long Island Hurricanes

Year	Impacts (state and SS category)[a]	Name[b]
1858	NY1 CT1 RI1 MA1	
1869	NY1 RI3 MA3 CT1	September Gale of 1869
1869	MA1 ME2	Saxby's Gale
1879	MA1	
1893	NY1 CT1	Midnight Storm
1894	NY1 RI1 CT1	
1896	RI1 MA1	
1924	MA1	
1934	NY1	
1938	NY3 CT3 RI3 MA2	Great New England Hurricane of 1938
1944	NY2 CT1 RI2 MA1	Great Atlantic Hurricane of 1944
1953	ME1	Carol
1954	NY3 CT3 RI3	Carol
1954	MA3 ME1	Edna
1960	NY3 CT2 RI2 MA1 NH1 ME1	Donna
1969	ME1	Gerda
1972	NY1 CT1	Agnes
1976	NY1	Belle
1985	NY3 CT2 NH2 ME1	Gloria
1991	NY2 CT2 RI2 MA2	Bob
2012	NY1	Sandy

NOTES: Includes all storms identified as having impacted any state in New England or New York as a hurricane. Significant and well-known storms that reached the region only as tropical storms, such as Floyd (1999) and Irene (2011) are not included.

[a] The data are shown in the same format as in the original HURDAT "trailer," which identified the maximum intensity of the storm (HR for all storms shown here), the states impacted and the Saffir–Simpson Scale category based on estimated sustained surface winds at each location. Only impacts in New England states and New York are shown. The information has been modified with updates resulting from the hurricane reanalysis project and as they appear in the newly published NHC's Chronological List of Continental U.S. Hurricanes.

[b] Names of storms in the 1800s from Ludlum (1963) and the same List of U.S. Hurricanes.

which were category 2 or 3 at some point during their journey through New England. There are also the very costly Agnes (1972) and Sandy (2012), and the famous "Perfect Storm" (1991). Floyd (1999) and Irene (2011) are also well remembered, but both had weakened to tropical storms by the time they arrived in the region. All of these systems caused significant localized damage, but none came close to matching the devastation and widespread damage of the 1938 New England Hurricane and its two historical counterparts.

The Previous Great New England Hurricanes: 1635 and 1815

The NHC's historical hurricane database (HURDAT/2) originally included storms dating to 1886 and was later expanded to 1851 after an exhaustive analysis of historical information (Chapter 1). To study storms occurring before then, one must look to other sources, but specific information on position and intensity for early storms is scarce. Existing historical resources are mostly qualitative in nature. The information is scattered in numerous personal journals, letters, newspapers, journal articles, and many other sources. Among these, two sources do stand out. Sidney Perley's 1891 *Historic Storms of New England* contains an exhaustive collection of natural phenomena experienced in the region since colonial times (not just hurricanes, blizzards, and other types of storms but also earthquakes, meteorites, and even northern lights and pollution events). The most complete and detailed information is contained in a 1963 American Meteorological Society monograph by David Ludlum, *Early American Hurricanes 1492–1870*. The work contains as detailed information as possible pertaining to hurricanes or suspected hurricanes from the days of the Spanish colonization of the Caribbean region through the creation of the agency that later became the U.S. Weather Bureau.

Only two other hurricanes have been of comparable strength and widespread devastation as the Great New England Hurricane of 1938: the Great Colonial Hurricane of 1635 and the Great September Gale of 1815. Using the few pieces of information available, Ludlum put together a remarkably detailed description of the Colonial Hurricane, especially for such an early storm. He writes:

> This was the great meteorological event of the colonial period in New England, coming only 15 years after the settlement of Plymouth Plantation and in the 5th year of the Massachusetts Bay Colony. It would be 180 years before another great hurricane of similar importance would strike the area with equal force. . . .

Perley also described the effects of the storm:

> The wind caused the tide to rise to a height unknown before. . . . An inconceivable number of trees were blown over or broken down. . . . It was universal,

> no part of the country being exempt from its injurious effects, which visibly remained for many years.[5]

Very few measurements from this storm are available, and the descriptions above are mostly based on the writings of two prominent colonists. John Winthrop, governor of the Massachusetts Bay Colony, kept a detailed day-to-day journal of happenings in the Boston area, and his writings prove him to have been a keen observer of the weather. His journals are the most detailed source of information about the conditions associated with the storm. William Bradford, governor of Plymouth Plantation, also wrote about the event. The most important portions of both men's statements are included and analyzed in detail in Ludlum's study. The information is mostly qualitative (except for a handful of measurements of water height in feet), but it allows the inference that, even though a much smaller population was affected by and reported the storm, its effects (specifically the extreme storm surge and massive loss of trees) were similar to those of the 1938 storm. The information is relatively more plentiful for the Gale of 1815, which was documented by an abundance of newspaper articles, a few scientific journals, and many town histories. Of this powerful storm, Ludlum writes:

> Few local histories fail to mention some incident or reminiscence to add to the chronicle of devastation and horror left by the most powerful atmospheric disturbance to lash the area since the Great Colonial Hurricane of 1635. And 123 more years would pass until September 1938 when another great storm of like dimensions and intensity would create such widespread havoc.

Ludlum goes on to describe the 1815 storm in great detail, using many sources to reconstruct the story. Perley additionally describes the 1815 season as "remarkable for the exceptionally violent and disastrous storms all along the Atlantic coast . . . [causing] great destruction of life and property on both land and sea." This seemingly infers that the 1815 hurricane season was particularly active. Yet, he singled out the September Gale by saying:

> The equinoctial gale of September, however, exceeded them all in violence, and caused greater and more general disaster than any that had preceded it, not that year only, but since the settlement of the country.

Besides these general descriptions, all the specific signs of a massive storm were there: the extreme "storm tide," the widespread loss of trees, and in this case, also the reports of salt water being carried by the wind tens of miles inland (just like it happened in 1938). An interesting aspect of this storm is that a timber salvage operation of sorts took place, foreshadowing the massive effort that would follow the 1938 storm:

> The loss in timber trees was exceedingly great, and in order to save as much as they possibly could from the ruins of their forests, the owners had the logs sawed into lumber with which they constructed houses, barns and other buildings. Probably New England never knew another season of such building activity as prevailed in 1817 and 1818, the logs having been sawed in the winter of 1815 and 1816 and the lumber seasoned during the following summer. This occurred in hundreds of towns and villages. . . .[6]

Sedimentary Evidence of Pre-Colonial Hurricanes

Historical accounts are not the only evidence of the extraordinary significance of these storms. As suggested in the discussion about the coastal changes brought on by the 1938 Hurricane (Chapter 8), a storm like this can have dramatic effects on the barrier islands outlining the coasts of Long Island and southern New England. The relentless crashing waves battering the beaches as the storm approaches with the powerful scouring by the storm surge overtopping them at the height of the hurricane can scrape off sand and gravel and breach the bars by cutting channels through them. The removed materials are then deposited in fan patterns on the bay side off the sandbars and on inland wetlands along the southern coasts of the region. Sediment accumulations in these locations thus consist of alternating layers of natural year-to-year deposition process materials and sand deposits washed out by powerful storm surges. For this to happen, the storm surge brought on by a particular storm combined with the natural tides must be high enough to overtop the barrier islands, suggesting that only the strongest storms will leave a sedimentary signature. The approximate age of deposited sediments can then be determined by using different archeological dating techniques to verify the signature of known storms and learn about prehistorical storms present in the record.[7] Careful analyses of sediment cores obtained at various locations along the southern coasts of eastern Long Island, Connecticut, Rhode Island, and Cape Cod suggest that the largest, thickest sand layers (coming from the most powerful storm surges) can easily be correlated to the well-known historical storms: 1938, 1815, and 1635.[8] The specific location where a core is drilled determines which hurricanes' sediments can be identified more easily, since each storm has a slightly different track. This is especially true for smaller hurricanes that, even if intense, have more localized effects and can only be detected in a smaller number of cores drilled near those locations. This was the case with Carol in 1954 and the 1869 category 3 storm, which appears to have been a remarkably small system. The sediments resulting from the largest (and most powerful) storms are more widespread and show up in cores all along the coast. Some uncertainties notwithstanding, there is good agreement between the historical and geological evidence about which storms were most significant for the region (1635, 1815, and 1938).[9]

The study also found sedimentary evidence of at least three other storms dating back approximately a thousand years: one during the first half of the 1400s (dated sometime between 1404–1446), one during the 1300s (1295–1407), and the earliest one during the first half of the 1100s (1100–1150). A simple average of the estimated six intense and devastating storms in the span of 900–1,000 years thus yields an approximate recurrence period of 150 years, consistent with our frequency discussions above.

Comparing the Great 1635, 1815, and 1938 Hurricanes

This book has covered much detail about the meteorological characteristics of the 1938 Hurricane and the environment in which it existed and moved when it approached New England. If similar characteristics can be discerned for the 1815 and 1635 storms, it might be possible to construct a simple model for the most intense and devastating New England hurricanes. Similar available measurements from the earlier storms are understandably scarce, but there is one piece of information that was more likely to be recorded than any other: the height of the water—storm surge plus tide for coastal locations and river height for inland locations. A 2006 study by retired NHC storm surge expert Brian Jarvinen took advantage of this fact.[10] He used available information (mostly Ludlum's analyses) to estimate the storm tracks and a first guess of storm intensity and size. The water height reports are often of the total height (or storm tide), which includes storm surge and astronomical tide; therefore, he also used a tide model to determine the actual storm surge represented by observations. He then ran the NHC's storm surge forecasting model, SLOSH (Sea, Lake, and Overland Surges from Hurricanes), and varied the size and intensity parameters until the storm surge matched the observations as well as possible. It immediately became obvious that all three of the storms had very similar characteristics (see Table 9.5): high intensity (as inferred from estimated central pressure and one-minute over-the-water winds, as calculated by SLOSH), a relatively small area of most intense winds,

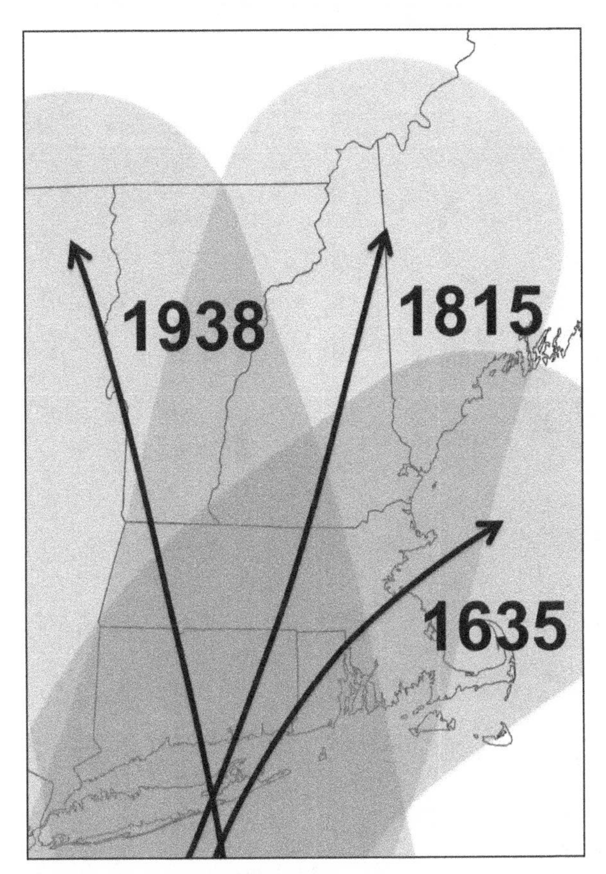

The three "big ones" (The Great Colonial Hurricane of 1635, The Great Gale of 1815, and The Great New England Hurricane of 1938) made landfall somewhere in Long Island followed by a second landfall in southern New England. They were intense, large storms whose effects extended well away from the center of their track, causing massive and widespread devastation from the coastal areas to deep into the forest of the White Mountains. The gray shading is not meant as an exact depiction but as a schematic representation of the large size of the storms and the extent of their associated damages.

a high translational speed, and a northward direction of motion. The results also included a high storm tide at Providence and Buzzards Bay, but this was the prescribed criterion that all other parameters were adjusted to produce, and therefore an expected result.

That all three of the systems approached the region from a southerly direction suggests that the large-scale meteorological conditions producing the air currents moving the storms must have been similar, at least in the broader sense of a possible deep upper-level trough of low pressure over

TABLE 9.5. Comparison of Storm Characteristics for the Most Intense New England Hurricanes

Storm	Landfall Location	Pressure (hPa)	Wind speed[a] (mph)	RMW[b] (mi)	Motion[c]	Storm Tide[d] Providence (ft)	Storm Tide[d] Buzzards Bay (ft)
The Great Colonial Hurricane of 1635	Long Island	938	132	35	SW/40	14[e]	>20.0[e]
	Connecticut	939	130	35	SW/40		
The Great September Gale of 1815	Long Island	956	122	30	SSW/47	14.4	15.9
	Connecticut	957	122	30	SSW/47		
The Great New England Hurricane of 1938	Long Island	941	132	30	S/40	15.8	14.1
	Connecticut	946	129	30	S/38		

SOURCE: SLOSH simulation results from Jarvinen (2006), *Storm Tides in Twelve Tropical Cyclones.*

[a] SLOSH one-minute wind over water (without frictional effects of land).

[b] Radius of maximum winds, defined as the distance from the center of a tropical cyclone to the location of the cyclone's maximum winds, normally found at the inner edge of the eyewall in a well-developed hurricane.

[c] Translational direction (defined as the direction from which the hurricane is coming) and speed (mph). The direction appears opposite to Jarvinen (2006), which instead listed the compass heading, but it represents the same direction of motion (generally from south to north).

[d] Storm tide includes storm surge and astronomical tide at the time of landfall.

[e] Storm surge rather than storm tide, since astronomical tide is not available for the 1635 storm.

North America with a large Bermuda High over the ocean (see Chapter 6). The high translational speed signifies that the steering currents were not only in the same direction but also of similar strength, meaning that the pressure features were intense enough to result in a large pressure difference between the two and hence a stronger pressure gradient force (see Chapter 4) "pushing" the steering winds. The storm motion speeds are also interesting in that they show that the 1938 Hurricane was not moving as fast as its reputation suggests (as was already determined in Chapter 6) and that other storms have moved as fast or faster as they came into New England. The small radius of maximum winds (RMW) would seem to clash with the fact that the widespread devastation produced by these storms needed a large wind radius, but both attributes are not necessarily exclusive of each other. All three storms were very likely transitioning into extratropical storms and expanding their wind shield as they were making landfall. At the same time, the maximum winds, which would have only been close to the center and enhanced by the large translation speed, would diminish immediately after landfall, hence the relatively small RMW.

Putting all this information together provides a simple model to describe the most intense and devastating New England hurricanes: they are fast, northward-moving storms that are still holding on to significant intensity in their core while transitioning into extratropical storms, hence enlarging the reach of their winds. This unlucky convergence of general characteristics, though infrequent, is clearly not unique to the 1938 Hurricane.

When It Happens Again

As evidenced by the long period between storms, the coming together of the meteorological and environmental factors necessary to create storms like that of 1938 (or 1815 or 1635) is uncommon. Neverthess, all evidence points to a storm as large, strong, fast-moving, and devastating as the Great New England Hurricane of 1938 making landfall in the region in the future. When that happens, some aspects of the storm would likely be similar to those of 1938, and some would be very different. Considering a historical storm in contemporary conditions is problematic. Even though the storm itself will most likely have the potential to cause the same type of weather conditions and effects, the people and the environment in the region have changed tremendously, and this evolution means that other aspects (such as communication, preparedness, relief efforts, and adjusted cost) will be very different.[11]

We must, of course, address the question of whether a storm similar enough to be called a "recurrence" has already made landfall in the North-

HURRICANE SANDY

Hurricane Sandy did not make landfall in the northeastern United States as such, but as a post-tropical system. It had fully undergone an extratropical transition by then even with its core winds still of hurricane strength (80 mph). Sandy's metamorphosis was not ordinary by any means. After the transition commenced as the storm gathered fronts and tapped into a new energy source, the core of the system maintained its tropical nature and even reintensified somewhat. Consequently, for a while, the system was described informally in the media as a hurricane inside a nor'easter. In reality, however, the storm should actually be classified as hybrid. The NHC does not distinguish between the tropical core and its surrounding extratropical storm environment when determining storm size and hybrid Sandy grew to be the largest storm ever observed over the Atlantic Ocean. After severely affecting parts of the Caribbean, the storm simultaneously weakened from a major to a minimum hurricane and grew to an extraordinary size. Its tropical storm-force wind field reached an astonishing 1,000 miles in diameter at its largest extent (the NHC estimated a maximum 870 nautical mile diameter). As a result, impacts were much more widespread than one would expect from a storm with the strength of a minimum hurricane, as Sandy was during landfall. Additionally, the interaction of the system with a complicated setting of upper-level steering currents caused an unusual track straight westward into New Jersey, in the process sparing New England the worst impacts. Lower Manhattan was left under water and boardwalks and homes along the New Jersey coast were destroyed by a catastrophic storm surge, but the large storm was also felt from Virginia north to Maine and west to Michigan. Sea level rose throughout the eastern seaboard, but also damaging surge pounded the

east. Hurricane Sandy's effects in 2012 in New Jersey and New York were costly, devastating and deadly, but how does it compare with the "big three?" Sandy was a very large storm, felt along almost the entire east coast and a large portion of the eastern half of the country. The areas that felt its most destructive impacts, however, were localized and not as widespread; New England was spared Sandy's worst impacts, unlike the situation in 1938. Further, its intensity at landfall was not that of a major hurricane. Superstorm Sandy, as it was dubbed, was a strange storm and in many ways defied our

Great Lakes, and a blizzard blinded West Virginia and parts of North Carolina with several feet of snow. This is the only time in history tropical and blizzard advisories were issued at the same time, and due to the same storm system.

More than meteorological peculiarities make Sandy interesting. The anticipated landfall as an extratropical cyclone meant that criteria for hurricane warnings would not be met. The NHC decided that it would not issue any of these alerts in advance of the storm for the region where landfall was expected to occur, as it might cause confusion when they were removed when the storm became extratropical. Unfortunately there was still confusion, but instead it concerned the nature of this storm and how dangerous it would be. It is not surprising, then, that the situation resulted in harsh criticism of the NHC's practices. To avoid the same situation in the future, the official definition of a hurricane warning and other advisories was revised on June 1, 2013 in order to allow for their issuing or continuation in association with post-tropical cyclones.

Sandy also arrived just a week before election day, when the nation would cast its votes for president of the United States. The storm disrupted campaign activities and changed the tone of pre-election conversations, and this fascinating intersection of politics and meteorology has already been and will continue to be highly examined by the media.

On a lighter note, the storm's approach just before Halloween informally earned it the nickname "Frankenstorm." Soon, however, it became known as "Superstorm Sandy," which is appropriate, given the extent and nature of its meteorological and social impacts.*

* Much more information about Hurricane Sandy, its meteorological history, and its impacts can be found in the official NHC tropical cyclone report on Sandy and in a very informative article in *Weatherwise* (Halverson and Rabenhorst 2013).

understanding and expectations. Accordingly, it deserves close examination in both its scientific and social nuances. Its impacts, however, as bad as they were, fell well short in comparison to those that a repeat of a storm like the 1938 Hurricane would have had.

When the next big New England hurricane comes, wind, storm surge, and precipitation potential may be similar to that experienced in 1938, but the details (where the strongest winds will hit during landfall, where the highest storm surge will happen, how the storm surge will be enhanced by the shape of the coastline, what widespread area will experience significant rainfall, etc.) will depend on the specific track of the storm. Other factors will also come into play.

The amount of inland flooding, for example, will depend on the capacity of the soil to absorb the hurricane's precipitation, which will itself depend on how much rain has recently fallen in the area, as well as how wet the preceding summer and early fall have been. Soil moisture will also affect the degree to which trees will be vulnerable to the wind. Inland flooding will depend on the infrastructure in place to control said flooding, such as dams, walls, and levees, many of which are new or have undergone significant renovation since the Hurricane. It is worth noting that evacuations for river flooding during a hurricane are not as well organized or at least not as clearly apparent to the public as coastal evacuation. This means that modern inland communities in river flood plains are in perhaps greater danger than the coastal population. In fact, before Hurricane Katrina (2005), the largest percentage of modern hurricane fatalities was due to inland flooding (as briefly noted in Chapter 6).

Besides the track specifics and coastal topography, the severity of the total "storm tide" will additionally be at the mercy of the combined timing of the arrival of the strongest inbound winds coming with the eyewall and the local astronomical tides, which will either add to or subtract from the water height the storm is capable of producing on its own. In other words, the specific environmental conditions encountered by a storm with the same characteristics as the 1938 Hurricane will determine in great part the intensity of the hazards experienced due to its arrival.

Worse Damages?

Changes to structure and infrastructure development make the damage of an equivalent storm potentially significantly more serious than it was in 1938. The coastal and river flood plain population is now larger. The year-round population is larger as well, and people have built more, and more costly, homes and other buildings (even after adjusting for inflation). The utilities infrastructure has also significantly expanded, with the addition of cable TV lines and cell phone towers. Many of these towers and poles could be knocked down, and many trees might fall over power and utility lines. Not only is today's population much more dependent on electricity but there are also more cars and more reliance on roads. Heavy traffic jams would be very likely during a hurricane evacuation and the subsequent returning home would be impaired by washed out and blocked roads. This would also have

MORE TREES!

Given the same wind, rainfall, and soil moisture conditions, even more tree damage could occur today than it did in 1938. The region was then both recovering from the excessive clearing and logging of the previous decades and shifting away from an agricultural economy. Many of those then sparsely forested areas are now covered by trees. According to the Forest Service's Forest Inventory and Analysis, for example, the state of New Hampshire, which experienced major blowdowns in 1938, is now nearly 85 percent covered by trees, up from a low of around 45 percent in the mid-1800s.

been the case in 1938, except that without advanced warning and the absence of evacuation, everyone was at home or at least in the general area when the storm arrived. Today, we might evacuate, but we would be leaving behind more; and massive coastal development and dependence on infrastructure make us vulnerable and recovery costly.

According to U.S. Census data (collected every 10 years), the total population of New Hampshire, Vermont, Massachusetts, Connecticut, Rhode Island, and the Long Island portion of New York was 12.2 million in 1940 and 20.7 million in 2010. These numbers represent a population increase of 69.7 percent. A simple proxy for coastal population could be achieved by using only the population of Rhode Island, Connecticut, and Long Island, which totals 7.0 million and 12.2 million for the same two decades. This represents a similar, although slightly higher, increase of 73.7 percent. In other words, the number of people living in the areas affected in 1938 is now approximately 70 percent larger, or 1.7 times what it was then. The census data also include housing information. The increase in coastal residential structures (also using total Rhode Island, Connecticut, and Long Island numbers) is even more impressive, with more than twice (2.4 times) as many homes and apartments as existed around the time of the Hurricane (an increase from 2.0 to 4.8 million "dwellings").

How Much Will It Cost?

Calculating the cost of a very similar hurricane under contemporary conditions is a complicated matter. An inflation adjustment is easy to apply, and as stated in Chapter 1, the 1938 estimated cost of $250 to $450 million would translate to approximately $4 to $7 billion in 2012 USD. Adjusting for the increase in population and development levels, however, can provide a more realistic idea of the potential cost of such a storm. There is no one way to do this, and thus researchers employ different methodologies. For example, a 2008 *Natural Hazards Review* article, ranked the costliest tropical cyclones of all time in two different ways.[12] The first method takes the storms' original cost adjusted by inflation (based on Bureau of Economic Analysis data), wealth per capita estimations (which take into account that people not only have more property and possessions but also that in many cases those have increased in real, noninflated value), and population. The total cost of the Hurricane obtained by using this method was $39.2 billion in 2005 USD. The second method used the same inflation adjustment, but the wealth calculation was per housing unit (nationally), and instead of population it took into account coastal county housing units as the third adjustment. This method resulted in a very similar $37.3 billion 2005 USD cost. Both results made the Great New England Hurricane number 6 in the Top 50 Damaging Storms listed in the study. It is also common for private risk management consulting companies to make these types of calculations (mostly for their insurance company customers) based on the storm data as well as various other economic, demographic, environmental, and proprietary databases and models. One such calculation, highlighted in the same article, resulted in an estimated $70 billion in total economic losses for the same storm in current wealth and population conditions. The Hurricane ranked as number 4 in their Top 10 Events list. Applying a simple inflation adjustment only, these three 2005 cost estimates would translate into $46.0, $43.8, and $82.1 billion in 2012, respectively.

The National Oceanic and Atmospheric Administration (NOAA) Technical Memorandum "The Deadliest, Costliest and Most Intense United States Tropical Cyclones" uses similar methodologies and includes various cost-ranking lists.[13] When simply adjusting for inflation, the Hurricane was ranked as number 19, with a cost of $6.1 billion in 2010 USD. When applying the "inflation, population and wealth normalization," the storm was ranked as number 6, with a cost of $41.1 billion in 2010 USD. When adjusted in this way, after so many years, the 1938 Hurricane remains the costliest New England hurricane. Until the arrival of Sandy (2012), it was also the costliest storm to have affected the northeastern United States.[14] The top five spots in the adjusted list belong to the 1926 Miami Hurricane, Katrina (2005), two Galveston hurricanes (1900 and 1915), and Andrew (1992). Sandy (2012) will most likely slide into position number 4 or 5 once the next update is completed.

A report on the storm by another consulting company, Risk Management Solutions (RMS), is worth discussing in some detail.[15] It covers the various aspects of storm damage and cost resulting individually from the wind, rainfall, and storm surge "footprints" of a storm with the same characteristics as the 1938 Hurricane, but in 2008 conditions. The estimated insured losses due to wind damage totaled $30 to $35 billion (two-thirds from residential losses and one-third from commercial losses). The insured cost due to the storm surge portion of the damage ranged from $6 to $10.5 billion. Inland flooding due to the excessive rainfall would be significantly reduced from what it was in 1938 in many places (thanks to the various flood control dams, pumping stations, and the like, which have since been erected, especially along the Connecticut River). An estimate of the cost of the same extent of flooding would therefore be an overestimate, but at least it would give a worst-case scenario ranging from $6 to $7.5 billion. Adding all loss estimates together translates into a range of $42 to $53 billion in 2008 USD in insured losses ($44.7 to $56.4 billion in 2012 USD). A rule of thumb commonly used by the NHC to estimate total economic loss is to double the insured loss. This would mean an estimated total economic loss of the order of $100 billion. However, when inland flooding represents a significant portion of the damages of a storm, as it did for the 1938 Hurricane, the doubling is known to often result in an underestimated cost.[16]

While calculating costs for a modern-day environment results in a wide range of estimates, it is clear that such a storm would result in major economic disaster. The 1938 Hurricane's characteristics, however, would not create the worst-case scenario, since the storm missed the highest population areas of New York City. A slightly different track could tell a very different and much worse story. Considering a database of possible category 3 storm tracks that could affect, not just New England, but also the New York coastline (including the New York City area and Long Island), the same study calculated the range of costs based only on wind damage.[17] The cost varied significantly, roughly from $1 to $150 billion, depending on the specific track and storm characteristics. The track corresponding to the 1938 Hurricane resulted in a cost of $35 billion in 2008 USD. The worst modeled damages resulted when the New York City area was under the right side of the eyewall (which as discussed in Chapter 6 is the most damaging part of the storm) or to its immediate east. Doubling the insured costs to estimate the total losses, the worst-case scenario of $300 billion in 2008 USD ($320 billion 2012 USD) is still an underestimate as it includes wind damage but not storm surge and inland flooding. Such damage would represent a financial catastrophe never before caused by a natural disaster in known history.[18] Sandy's path very closely aligns with a potential worst-case scenario. It produced immense losses, initially estimated to cost about $71 billion, mostly from coastal damages. However, it was not as intense as a major hurricane during landfall and its significant damages did not extend into interior New York and New England as with the "Great" historical storms (1635, 1815, and 1938); in that way it fell well short of the actual worst-case scenario and maximum potential cost.

The Good News

The tools and practices that have since been developed, on the other hand, paint a more reassuring picture. Meteorological satellites monitor incoming hurricanes from formation through dissipation, transition, and landfall. The position of the eye of a storm nearing the coast would additionally be detectable by National Weather Service (NWS) radar due to the intense eyewall rainfall. NOAA and Air Force hurricane hunter missions would fly through the storm to take measurements, and as much information as possible about the storm's behavior and surrounding meteorological conditions would be fed into the various global and hurricane-specific forecast models to predict its future behavior. Therefore, assuming no loss of monitoring or forecasting capability, an unannounced hurricane with an unexpected storm surge killing hundreds or more could not happen again. In fact, the entire nation would likely know of the possibility of a New England strike a few days in advance, just as with the expected arrival of Hurricane Irene in New England in 2011 and Sandy in New Jersey and New York in 2012.

Even though there are still uncertainties, and accuracy varies from storm to storm, the average error in forecasting the track of a hurricane has decreased systematically and significantly over the past few decades. For example, in 1970, the average one-day track forecast error was 140 nautical miles (n mi); in 1990 it was 100 n mi and in 2010 it was 40 n mi.[19] A four-day forecast is currently as good as a two-day forecast was in 1995, and as good as a one-day forecast in 1970. The intensity of a storm, on the other hand, is dependent not only on more easily forecast larger-scale conditions but also on internal storm mechanisms and small-scale effects that are not as easy to anticipate by forecast models let alone forecasters themselves. Consequently, rapid changes in intensity are still not as accurately forecasted, and it is not unusual for a storm that is expected to maintain its intensity to suddenly weaken or conversely, and more seriously, for one that is expected to weaken or maintain its intensity to undergo what is known as rapid intensification. Indeed, intensity forecast errors have barely improved over the last 20 years.

It is feasible, then, that significant uncertainty about the intensity changes of this hypothetically serious hurricane nearing New England would be in play as late as the day of landfall. Hurricane Irene (2011) provides an excellent example. Its track into Long Island and New England was accurately forecasted five days in advance of landfall. The forecast intensity of the storm for its New England approach and arrival, on the other hand, was consistently too high.[20]

Communications

One of the greatest differences between what unfolded in 1938 and what could unfold today is that now almost everyone within the region would be alerted about a dangerous storm approaching. In 1938, communication was not instantaneous. A hurricane warning would first reach local authorities, who would convey the announcement to the general public. This could take hours. Common media of the time—newspapers, announcements posted at city hall or post offices, and word of mouth—were slow by nature, and warning of an incoming storm only one or two hours in advance would have been almost impossible. Radio, telephone, and telegraph were faster, but they could not have reached an entire regional population, especially one that was expecting nothing more than a gale.

Today's population is highly connected. While newspapers, radio, and television would certainly relay information about the incoming storm, many other outlets are now more common and efficient. Twenty-four-hour cable stations dedicated to news, The Weather Channel, and numerous Internet sites make information accessible to the public in real time. Social networks buzz with news of every oncoming storm. Sandy (2012) provided the opportunity to observe such communication and readiness in action, and it made clear that social networks are especially dominant both in serving to inform the public and in reporting storm-related emergencies and damages.[21]

Even more importantly, the original source of official information, the NHC, posts all advisories with storm information, discussions including track and intensity forecasts and forecast rationale, and storm watches and warnings online for all to access. The NHC also coordinates closely with emergency management officials and with NWS Forecast Offices, which in turn issue forecasts of the expected weather conditions for their specific locations and further coordinate with local emergency management offices to prepare the public and make ready shelters and other resources required for safety and survivability in the aftermath. This life-saving information is then communicated to the public via numerous media outlets.

Clearly, lack of information would not be the point of failure if a major storm were approaching today. During and after the storm, however, would be a very different story because without electricity, only those who were prepared with alternative resources, such as battery-powered radios, would be able to stay informed directly.

Emergency Planning and Relief

Sophisticated preparedness plans have developed over the years and many layers of emergency measures are now in place. From national to state to local levels, each location has a response plan (or more than one) ready to be put into action at a moment's notice. The Great New England Hurricane of 1938 is often used as an example of a worst-case scenario for planning purposes, especially within the affected regions. At some point prior to a storm of this magnitude, evacuation orders (possibly voluntary at first) would most likely be given for those areas expected to flood with life-threatening storm surge. Evacuation routes would be used and contingency plans (such as using inbound lanes for outbound traffic) would be activated. Areas to be evacuated would depend on the magnitude of the expected coastal water rise, which itself, as discussed, would depend on the strength and size of the storm, the specific landfall location, and the timing of landfall. Emergency management agencies and personnel would work closely with the NHC and local NWS offices and their forecasters to make informed decisions with as much advance notice as possible.

While there are more roads, more utility infrastructure, and more houses and buildings, many of them are better constructed, sturdier, and less vulnerable than they were many decades ago. There are also more sophisticated cleanup tools and easier accessibility to heavy equipment for the task. International agreements with Canada are also in place, and resources can be obtained from Quebec and the Maritime Provinces to aid in northeastern natural disaster recovery.[22] Additionally, roles for various relief agencies are now clearly defined and organized.

The Bottom Line

In the end, when it happens again, the overall impact of such a storm will depend on the balance among a few simple factors.

As long as the current monitoring, forecasting, and warning communication capabilities are operational, an unannounced landfall in New England will never happen again. Emergency preparedness and relief protocols are in place, ready for when landfall does occur. Today's northeasterners, even if still inexperienced in dealing with major hurricanes, understand their nature and potential hazards much better than their counterparts decades ago. On the other hand, infrastructure, buildings, and population have significantly increased. This means that extreme and expensive damages would occur, but a death toll of the same magnitude as in 1938 would be highly unlikely.

Remaining Scars and Enduring Legacy

Three quarters of a century after the Hurricane, there seem to be no obvious visible environmental scars. It is much easier to see the localized effects of more recent storms, such as the roads and bridges repaired and rebuilt after Hurricane Irene's 2011 flooding washed them out, or broken and fallen trees blown down by an ice storm or even by a much smaller thunderstorm carrying severely strong winds. To those who look carefully, however, signs of the Great New England Hurricane of 1938 can still be found.

Dramatic coastal modification (described in Chapter 8) resulted in new beach inlets (such as the Shinnecock Inlet in Long Island) and new islands (such as Sandy Point in Rhode Island). Most signs, however, are subtler. There has been enough time now for the fallen forest to regrow. New trees have replaced fallen ones, and patches of what is called pit and mound topography can be found deep in the woods. When a large tree is uprooted and blown down, its roots tear a portion of the soil away, leaving behind a bowl-shaped depression or "pit." The tree material decays over time, leaving behind a "mound" of dirt where the roots of the fallen tree were once dangling in the air. These are natural structures found in mature forests, but a catastrophic blowdown event (such as the 1938 Hurricane) leaves behind extensive clusters rather than individual pits and mounds. Today, they can be found throughout the White Mountains, together with trees dating to just under 75 years old (meaning they were the first growth after the Hurricane-induced blowdown).[23]

The Hurricane left its historical stamp everywhere in the region. In the words of *Yankee Magazine* Editor-in-Chief Judson D. Hale:

> . . . there's hardly a New England town in existence today that does not possess, in some form or another, tangible, physical proof that a certain blockbuster of a hurricane came roaring through New England one September day many long years ago. No other hurricane can claim the same.[24]

Water marks on buildings; plaques showcasing said marks or commemorating a structure or a bridge (especially historical covered bridges that did not survive the storm) or a memorable tree taken down; pictures inside historical buildings, hotels, and churches showing their damage: these can all be found throughout New England. Hundreds of historical societies serve as repositories of photographs and accounts from the storm's passage and its effects on their towns. Furniture built with wood fallen during the Hurricane, and marked as "made with hurricane wood," can still be purchased in antique shops and auctions. Souvenirs issued by the city of Hartford, Connecticut, to commemorate the relief efforts can easily be found and purchased online. There are also numerous newspaper articles, private picture collections, witness accounts, fictional stories, and even theatrical plays and musicals loosely or closely based on the events surrounding the storm. As an example, a recently debuted original community theater play commemorating the 250th anniversary of Plymouth, New Hampshire, highlights both the 1936 flood and the 1938 Hurricane as locally significant events happening on the eve of World War II with a song named "The Water Rose High."[25] About the Hurricane, it says:

> The wind began to blow and blow, to seal poor Plymouth's fate
> A hurricane was passing through, all they could do was wait
> It tore the limbs off of the trees, and damaged all with glee
> When morning dawned how sad it was, disaster for all to see
> The winds, they blew high
> The winds, they blew low
> The hurricane dealt us a fateful blow.

The significance of the Great New England Hurricane goes well beyond relics, witness accounts, pictures, letters, newspaper articles, and hurricane wood furniture. Of great meteorological importance is the fact that tropical scientists were able to observe for the first time the now-well-known extratropical transition of hurricanes. The storm also taught forecasters and the public that New England is not immune to storms originating in the tropics. The Hurricane came at a time when meteorology and forecasting in the United States were on the verge of decades of vast modernization and technological advance and, even if it did not directly propel the changes, it provided a compelling validation of the need for change. The Hurricane also marked a turning point in the way in which the federal government deals with emergency relief, for the first time taking a direct and massive role.

The recovery from the storm's aftermath served as a testament to the resilience of New England and Long Island, a fact boasted by numerous articles

The Historic Market House building (est. 1773) in downtown Providence, Rhode Island, features two detailed plaques marking the height of the water line for the two great New England storms through which it has lived. The water reached approximately two feet higher in 1938 than in 1815. The respective heights (13 feet 8½ inches and 11 feet 9¼ inches) are above mean high water (the average high tide) for that location.

The city of Hartford, Connecticut, assembled "Official Relief Relics of the Hurricane and Flood" by attaching brass-plated commemorative disks to sections of fallen tree limbs of different types and sizes.

WORKS OF FICTION INSPIRED BY THE HURRICANE

The House on Salt Hay Road by Carin Clevidence, 2010
Oliver's Surprise by Carol Newman Cronn, 2009
Moon Tide: A Novel by Clifton Tripp, 2003
Castle Ugly: A Love Story by Mary Ellen Barrett, 1966
Country Place by Anne Petry, 1947

AUTOBIOGRAPHICAL STORIES

Tales of a Farmboy by Clarence J. Salmon and Donna L. Salmon, 2007
26 Fairmount Avenue by Tomie dePaola, 1999

and reports immediately following the storm. Chairman of the American Red Cross Norman H. Davis stated:

> I have been deeply impressed with the courage and self-reliance evident everywhere, and particularly among those who have suffered most.[26]

Each individual town was in the end both scarred and proud of surviving the storm. As the song continues:

> But the people rose up, reclaimed their land
> Cleared all the trees, patched up the bandstand
> Despite all the work, despite all the fear
> Plymouth would go on, to live another year.

Today, the story of the storm fascinates those who seek to understand its scientific and historical complexities: the terrible beauty of a meteorological monster, met by a perfect confluence of circumstances, and resulting in astonishing levels of devastation. Only a handful of times in the history of the United States has a natural disaster brought such tragedy and destruction to an entire region, but Long Island and New England gracefully recovered. Lives, homes, roads, and bridges were rebuilt, the coastline settled into a new equilibrium, and forests regrew. The Great New England Hurricane

1938 earned a significant place in history with its complex and lasting social and meteorological legacy, from which much continues to be learned. So many decades later, it remains the storm to which all other New England hurricanes are compared.

In Providence, Rhode Island, history is palpable. The old and the new co-exist within the nearly 400-year-old city, which surely all too often is taken by storm, as it has been visited by many of the most significant New England hurricanes. The Great New England Hurricane of 1938 was the most destructive, but the Great Gale of 1815 and a little more recently Carol in 1954 also paid costly visits to the city.

Downtown, visitors can stroll among buildings that survived the Hurricane, many of which display plaques indicating where high-water marks were left after the storm waters subsided. Inside the lobby of the beautiful Biltmore Hotel, for instance, the plaque rests high above eye level. The same is true for those markers outside the historic 1773 Market House Building. Although many will walk past without noticing the plaques, a brief pause and reflection can both amaze and horrify those who consider how rough and high the waters that inundated the city were, the panic and confusion people must have felt as this water rushed in, and the sadness that the unimaginable loss left on its wake. It is also clear, however, standing there where so many lost their lives, that life does go on, cities rebuild, and progress continues, as evidenced by the seemingly unscathed buildings and liveliness on the streets today. Other hurricanes will certainly blow through, and flooding and destruction might once again come—and the city will once again recover.

Hundreds of other picturesque cities and towns similarly suffered devastating losses from the 1938 Hurricane, some greater than others, but they all have gracefully recovered. Nearly a century after the Great Hurricane, beautiful, historic New England is still here, and we are all thankful for that.

10
THE HURRICANE, REVISITED

There is still more to learn and more to say about the Great New England Hurricane of 1938, which is a remarkable thing, given that it struck so long ago, albeit with exceptional force. During the five years since the publication of the first edition of this book, updated forecasting practices, definitions, studies, tools, and interesting pieces of information have surfaced. This chapter covers a selection of these topics, and the book website www.takenbystorm1938.com contains additional pictures and other materials to supplement the information.

New Official Practices

The NWS/NHC continues to revise how it monitors, forecasts, warns, and communicates about hurricanes. It does so by taking advantage of new science, analysis tools, and modern communication platforms, and in response to issues of safety and clarity as they are exposed. The ability to continue a hurricane warning in threatening conditions even if a storm is no longer of tropical nature (an adjustment made following the communication issues experienced with Hurricane Sandy in 2012, as described in Chapter 9) and the addition of "post-tropical cyclone" terminology (described in Chapter 7) to the roster of hurricane-related definitions are examples of such revisions. As the public is made aware of threatening tropical disturbances earlier and earlier in the storms' development, and as people are flooded with information from multiple sources, it has become important to issue *official* information as early as possible. Even though the NHC was able to do this if deemed appropriate for a specific storm, its operational use of the new classification of "potential tropical cyclone" during the 2017 hurricane season formalized the practice by triggering the initiation of forecasting products, discussions, public advisories, and graphics earlier than in the past.

Operational directives, such as the criteria for issuing watches, warnings, and other advisories, are mandated by a series of National Weather Service official rules, and the changing of any of them requires an extensive approval process that includes the opportunity for public comment and internal committee review. The National Weather Service instruction 10-604 (NWS 10-604, newly titled "Tropical Cyclone Names and Definitions"), contains classifications, criteria, and other information that must be used in official NHC advisories and discussions. As of this writing, it was last updated on April 21, 2017, when a variety of significant revisions were made.* The current 10-604 directive now defines "potential tropical cyclone" as "a disturbance that is not yet a tropical cyclone but which poses the threat of bringing tropical storm or hurricane conditions to land areas within 48 hours" and revises the definitions of watches and warnings (for both hurricanes and tropical storms) to allow their use for potential tropical cyclones (in addition to ongoing tropical cyclones and post-tropical cyclones, as was already the practice). In this way, there can be a consistent message from the moment a disturbance shows potential for development until well after a hurricane dissipates, without unnatural breaks in the communication of hazardous conditions.

Perhaps the most significant additions made during the 2017 season are storm surge watches and warnings, which, as directed in NWS 10-604, can now be issued in advance of an expected hurricane landfall. This is the natural culmination of a multi-year evolution of the way in which the NHC deals with the forecasting and communication of storm surge possibilities. Chapter 1 discusses why, after many decades in use, the Saffir–Simpson scale no longer specifies storm surge as part of the characteristics of each hurricane category. In short, the change was made to avoid misleading the public, since even though the information correctly represented what might happen for an idealized average storm, many other factors besides wind speed affect

* The most-recently updated document containing the current directives and related information can be obtained from the NWS Directives System at http://www.nws.noaa.gov/directives/sym/pd01006004curr.pdf

the intensity of the surge accompanying a specific hurricane. The chapter references the possibility of a separate scale for storm surge that the NHC rejected in favor of including clear communication about possible flooding due to storm surge (and other storm hazards) in the official public advisories. This has now been successfully done for a few years. As newly defined, a storm surge warning

> refers to the danger of life-threatening inundation from rising water moving inland from the shoreline somewhere within the specified area, generally within 36 hours, in association with an ongoing or potential tropical cyclone, a subtropical cyclone or a post-tropical cyclone. It may be issued for locations not expected to receive life-threatening inundation, but which could potentially be isolated by inundation in adjacent areas.

Inundation is defined as "the total water level that occurs on normally dry ground, expressed in terms of height of water above ground level," and for a given storm and predicted conditions leading to storm surge, it is specific to each location. Storm surge watches are issued within 48 hours or earlier if necessary for the safety of evacuations and other storm preparations. As stated by the NWS in their announcement of the new advisories,* they are "issued to highlight areas along the Gulf and Atlantic coasts of the continental United States that have a significant risk of life-threatening inundation" and they are determined by "a collaborative process between the NHC and local NWS Weather Forecast Offices" using the SLOSH model (described in Chapter 9), the latest NHC official forecast and the typical errors associated with such forecasts, together with forecaster confidence for the current storm, continuity from advisory to advisory, and other factors deemed relevant for each particular case.

The Train, the Mountain, and the Hurricane

The Mount Washington Observatory has kept a continuous record of weather observations since 1932.† The morning after the Hurricane, the weather

---

* NOAA press release dated January 23, 2017.

† The Mount Washington Observatory is a non-profit scientific organization. The current observatory began keeping records in 1932 (even through private ownership of the land), but the U.S. Signal Service, the predecessor of the Weather Bureau, maintained a station at the summit from 1870 to 1892. The observatory originally occupied the Stage Office, a nearby building notable for being chained to the ground.

observer stationed at the summit of the mountain (which, as described in Chapter 6, is the highest peak in the northeast United States) added a brief report to the official observations:

> Violent winds continued with lessening force throughout [the] night. Many trees blown down on all sides of mountain. Carriage Road closed by fallen trees. New slides caused by high winds in Burke's [Burt?] and Ammonoosuc ravines with many trees torn up. Railroad trestle ripped up from below sky-line to just above halfway.

The mentioned railroad is the historic Mount Washington Cog Railway,* which climbs for three miles along the west side of the mountain through a terrain with average inclination of 25 degrees, and as high as 47 degrees at its steepest. In order to repair the extensive damage, the owner at the time, Colonel Henry Teague, obtained a loan of $60,000 (over $1 million 2017 USD) from Dartmouth College. Teague owned not just the railway and the lands through which it runs but also the land at the summit. He died in 1951, and, according to the historical summary provided by the railway, he "left The Cog and all of its property to Dartmouth College in gratitude of their financial assistance after the 1938 Hurricane." Dartmouth sold the land to the state of New Hampshire in 1964 to use as a park and leased the rest until 2008, when it finally sold the remaining property to the state. Today, the Mount Washington State Park occupies the summit (60.3 acres), and its visitor center, which includes a large viewing deck and a museum, and houses the Mount Washington Observatory (which continues to be staffed by weather observers 24 hours a day), provides refuge from the harsh, often extreme weather conditions for visitors who venture the arduous hike or the terrifying road to the top.† It seems that if not for the 1938 Hurricane, this land might have stayed in private hands and never transform into the popular and accessible destination it is today.

---

* The Mount Washington Cog Railway, the first mountain-climbing railway in the world, is designated as a National Historic Engineering Landmark. It uses a series of gears with teeth that engage on rail slots in order to work against gravity to haul the engine and cars up the steep incline. It began operations in 1868 and is still run as a tourist attraction.

† The Mount Washington Auto Road, known during the time of the Hurricane as the Carriage Road (mentioned in the observer report) is a 7.6-mile toll road that climbs 4,618 feet to the summit of Mount Washington. It opened to the public in 1861.

Precipitation Observations

Chapters 6 and 7 contain a detailed examination of the precipitation preceding and accompanying the Hurricane. This discussion focuses on the numerous rainfall observations included in the UGSG report "Hurricane Floods of September 1938" and puts them in the context of the previous discussion. The nearly 750 precipitation observations encompass a large portion of the northeastern United States: New Jersey; New York (including Long Island, New York City, and upstate New York); southern New England (Connecticut, Rhode Island, and Massachusetts); parts of northern New England (New Hampshire and Vermont); and portions of southern Quebec, Canada. While some rain certainly also fell in Pennsylvania and Maine (and significant portions of both states naturally show up in any maps drawn to visualize the reported data), the report does not include observations for any location in either state. The book's website contains the complete precipitation data (daily totals for September 12–22 for each station) and a variety of additional maps.

The daily totals clearly show the significant multi-day rainstorm that preceded the Hurricane (and lasted from the 17th to the 20th).* The precipitation directly due to the Hurricane fell on the 21st only, starting around noon in Long Island and ending around midnight in the northernmost locations, and it was followed by dry conditions. Observations for the 21st and 22nd completely capture the Hurricane precipitation. The coarseness of the daily observations, together with timing uncertainties (similar to those discussed in the aforementioned chapters), mean the data do not show a clean break

---

* This preceding rainfall, which likely is, or at least contained, what we today define as a PRE (Predecessor Rainfall Event, as discussed in Chapters 6 and 7), might also be associated with a phenomenon that has recently become more prominently known. An *atmospheric river* is defined by the AMS glossary as

> a long, narrow, and transient corridor of strong horizontal water vapor transport that is typically associated with a low-level jet stream ahead of the cold front of an extratropical cyclone. The water vapor in atmospheric rivers is supplied by tropical and/or extratropical moisture sources. Atmospheric rivers frequently lead to heavy precipitation where they are forced upward—for example, by mountains. (www.glossary.ametsoc.org)

The moisture graphs drawn by Pierce (1939) show such long and narrow corridor, which together with the large precipitation amounts suggests the possibility of an atmospheric river.

between the two events. It is well established, however, that there was such a break late on the 20th and early on the 21st before the Hurricane precipitation began. Additionally, some of the 24-hour rain attributed to the 21st might have occurred on the 20th (depending on when the observation was taken), providing an unnatural contribution to the Hurricane rain. Since the flooding, landslides, and other related effects were a result of both events together rather than one or the other, it is appropriate and convenient to examine the much less uncertain total precipitation.

The following map is a geographical information systems (GIS) visualization of the total precipitation for September 17–22, 1938, based on an inverse distance weighed (IDW) interpolation for 748 stations. Such interpolation is a smart guess of what the values would be for each location even if no data were available. This information can then be used to draw contours or shade regions for appropriate pre-determined values. In this case, precipitation totals above 1, 3, 5, 7, 10, and 15 inches are shaded to show the overall distribution. Notable are two small pockets of total precipitation above 15 inches in central Connecticut and central Massachusetts (this one a little hard to see, since it only comes from one station). The distribution also shows a broad slanted spine of totals over 10 inches that starts spottily over eastern New Jersey and Long Island and continues solidly through Connecticut, central Massachusetts, and into southern New Hampshire. An even broader but also southwest-to-northeast-oriented swath of seven inches almost fully covers New Jersey and extends into the southeastern portion of New York state and southern Vermont and well into New Hampshire. Totals of at least five inches extend farther into northern New Hampshire, central Vermont, and large portions of New York, with three inches reaching into Canada and halfway west through New York, and at least one inch covering the entire region. The total precipitation diagram in Chapter 7 was based on a qualitative examination of sample locations at each state (from data available from the National Climactic Data Center, now part of the National Centers for Environmental Information). When compared to the observations map, it does well with the overall shape of the distribution. It also represents the region with the totals higher than 10 inches very well, but severely underestimates the area that received at least five inches. Redefining the one- and five-inch contours as five and seven, respectively, would provide the best representation of the USGS observations. The NOAA precipitation map for the Hurricane (reproduced and discussed in Chapter 6) includes too many days to show only the Hurricane precipitation but too few to include both events; it also does well qualitatively when compared with the 17–22 total.

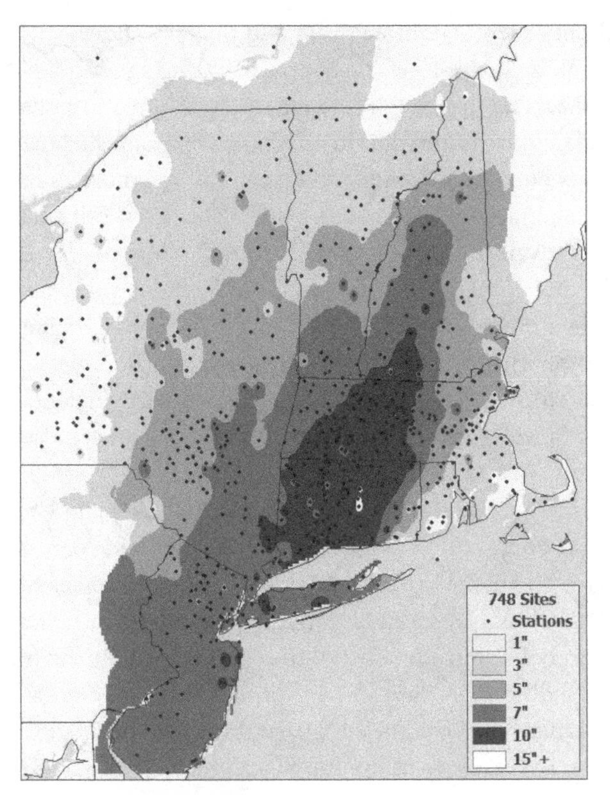

Precipitation observed during the Great New England Hurricane and its predecessor rain event as reported by the U.S. Geological Survey.

The Meteorology of the Storm

The meteorological conditions that allowed the Hurricane to come straight toward New England and with such strength are well understood. As stated in Chapter 6, there was a large and strong southward dip in the jet stream that steered the storm northward very fast. There were not enough upper-air observations at the time to directly observe this steering flow, but the signs are clear. Our previous most complete reconstruction of global atmospheric conditions (model-produced gridded values of variables such as temperature, wind speed, and moisture) was provided by the NCEP/NCAR Reanalysis Project and only goes back to 1948. Available more recently, however, the 20th Century Reanalysis goes as far back as the late 1800s.* A quick

* The 20th Century Reanalysis uses historical surface observations to recreate a 6-hourly, four-dimensional global atmospheric dataset starting in 1871, with the goal of placing current atmospheric circulation patterns into a historical perspective.

review of upper-level winds equivalent to the map shown in Chapter 6 (a composite of the upper-level winds occurring during several New England hurricanes) reveals the trough (the dip in the jet stream) might have been somewhat stronger during the Hurricane, which is not surprising since the storms that were used to compose the map did not move as fast. The 20th Century Reanalysis version is also oriented in a more north–south direction (or even slightly northwest–southeast direction, a hint of what is known as a negatively tilted trough) than the composite version, which would bring the storm straight north as it happened in 1938. The synchronizing of a feature in the atmospheric flow pattern with an incoming major hurricane in such a way that the storm will be steered straight toward New England is extremely rare (which is why such intense landfalls in the region are also rare, about once a century or two, as described in Chapter 9).

Robert Hart, professor of meteorology at Florida State University, has performed a variety of tests using Weather Research and Forecasting (WRF) Model ensembles of the storm based on the conditions specified by the 20th Century Reanalysis. An ensemble forecast is a numerical prediction method where instead of a single forecast of the expected conditions (or simulated historical conditions, in this case), the product is a set of forecasts that represent the range of possibilities. He used these simulations to assess information such as the rarity of the storm, its predictability, and the range of possible structures and timing of transition.

One of the greatest uncertainties about the Great New England Hurricane is the timing of its extratropical transition (as discussed in Chapters 1 and 7), which regardless of such uncertainty resulted in the worst-case-scenario impacts. It made landfall as a major hurricane but at the same time, though weakening, it was also expanding into a much larger system that could affect a much larger portion of the region. The storm's impacts reflect major hurricane winds and storm surge at the coast and immediate vicinity and extratropical cyclone wind and precipitation intensity and coverage throughout New England. In the words of the reanalysis project, the storm was "transitioning" as it made landfall. Even for a modern hurricane, with all the available data and analysis tools, it is easiest to know when a storm is still purely tropical and when it has fully transitioned, but it is much harder to identify the transition itself. In fact, there is still no operational definition of extratropical transition in any shape or form. The current determination is made subjectively, yet of course thoughtfully (and using all the available data), by the hurricane specialists at the National Hurricane Center. When the data are filed at the end of the season (and become part of the HURDAT2 records), there is an abrupt change in classification, a flip of the switch if you will, from tropical to extratropical in the storm information, giving the impression it happens instantaneously. This is not the case, and this is well known to tropical experts. As evidenced by a search of the scientific literature from the past couple of decades, however, the complexity of this transition and the questions remaining to be answered are outstanding. It is perhaps inevitable that an official definition for categorizing the transition objectively will be established in the near future, but exactly how this might be done is still an open question. It is clear that with it, some sense of the onset and completion of the transition will also be needed. In fact, all scientific studies on the topic define both in one way or another.

Professor Hart also developed what are known as storm "phase space diagrams" (famous among the tropical meteorology community).* They diagnose the temperature structure and symmetry of a storm at each stage of development and are the most useful tools to characterize tell-tale structural signs that allow an objective determination of the nature of the storm and its transition timeline. The most common transition path is for tropical cyclones to start the transition by a loss of symmetry and complete it by developing a cold core (fully transforming into extratropical cyclones). The full transition can take a day or so. The phase space diagrams for the ensemble forecast tracks and structures resulting from Hart's 1938 Hurricane WRF simulations show similar evolution from tropical to hybrid to extratropical structure, but they also show large variability in timing and exact core structure toward the end of the life cycle of the storm.

Deeper into The Forest

One of the goals for this project was not only to focus on the meteorology, but also to explore connections with other sciences. The forest, which was such a significant piece of the story of the Hurricane, served as one of the most important sources of interdisciplinary consideration. Topics in forest dynamics and ecology, and the various forest effects of the storm, appear in Chapter 7. Half a chapter, however, could never do justice to the complexities of how a hurricane of this magnitude affects such an extensive forested region, or even to the basics of how a forest ecosystem works. A book published in 2016 does just that. *Thirty-Eight: The Hurricane that Transformed New England* by Stephen Long beautifully and clearly explains everything

* See, for example, the *Monthly Weather Review* article "A Cyclone Phase Space Derived from Thermal Wind and Thermal Asymmetry" by Hart (2003).

and anything related to the storm's connections to the forest at a level accessible to a general audience. It covers forest history, ecology, and forestry, as well as tree farming and maple syrup production, land use, and other related topics, through the stories of people and places and the related processes and science. It also covers in great detail the forest effects of the storm and the massive cleanup and salvage operations that followed. It delves into the intricacies of forest dynamics, and about how and why the Hurricane affected some trees and some patches of forest more than others, and explains the signs of the storm still observable these many decades later.

Long describes in detail the numerous and uneven patches that pockmarked the massive footprint of 15 million acres of New England forest damage. A great majority were on the smaller end of the spectrum, as small as about one acre. The largest areas of continuous damage were as large as 60 acres but also much less plentiful. All through the region, unaffected or lightly affected patches of forest coexisted with the damaged areas. He also describes how the direction of tree-fall in a catastrophic large-scale event such as the Hurricane depends not only, as expected, on the direction from which the wind is blowing, but also on the direction in which gravity pulls the trees down. In steep terrain, the downward pull normally wins the battle, and the direction of fall matches the direction of the downward slope. In flat terrain, on the other hand, wind alone determines the direction (unless there are other considerations, such as blockage by other trees or structures), since in this case gravity does not favor any specific direction of fall. In between, both forces contribute,* but it is safe to assume wind direction wins in gentler terrain.

Long also describes remaining scars of the Hurricane that can be observed on the trees and the forest terrain. Conifers and hardwoods, for example, react differently to being knocked down. In simple terms, the conifers show a bend in the trunk at the point where the tree shifted growth to match the new upward direction, while the hardwoods show elbows where growth stopped for the original trunk and a new branch took over as the trunk. Together with aging techniques, such as coring and counting of tree rings, one can verify how long ago the traumatic events occurred. The pit and mound topography (described briefly in Chapter 9) is in fact much more informative than the simple fact that its widespread presence evidences a

---

* One can visualize the setup as a vector addition problem, with one arrow pointing downward along the slope and another arrow pointing to the side in the direction from which the wind blows.

large forest disturbance. It also shows the direction of fall: the mound, being the remnant of the tree material that decayed over time, points to the direction of fall like an arrow with origin in the pit (which is the depression left in the ground by the uprooting of the tree). This means that if the mound is to the northwest of the pit, the tree was knocked down by a wind blowing from the southeast, a common wind direction in New England during hurricanes. East and southeast winds were evidenced by a significant portion of the damage during the Hurricane, but some trees deviated from this pattern, matching the downslope direction rather than that of the wind. He also describes how the normal year-to-year leaf fall and decomposition and other forest floor accumulations are not enough to fill in the pit depressions. In fact, some pits are large enough to mostly or fully immerse a person below the surrounding ground level.

One of the most interesting updates coming from *Thirty-Eight* is the questioning of the assumption that the widespread tree damage resulted from the supersaturated ground (after days of extreme rainfall) not providing enough of an anchor for the trees as the strong winds blew. At least two experts independently expressed to Long they did not believe this to be the case. The argument is that root breakage could be clearly observed (as can be seen in photos), rather than the roots completely slipping out of the soft ground. One of them expressed that landslides would have occurred if the common theory was correct. Another mentioned that some terrains (for example, sandy soils) actually increase cohesion when moisture is added. Long used pictures to verify breakage occurred where the roots tapered down to the width of a human wrist. It is premature, however, to outright reject the common wisdom. The fact is that landslides did occur, with thousands reported, and the terrain types vary significantly throughout the region. The truth, as always, is likely more complicated: not all or nothing, but somewhere in between. Perhaps the roots would have held longer or they would have broken at larger widths if the ground provided a stronger anchor. Just like the forest devastation was highly uneven, the levels of soil saturation were also likely uneven, with lower and less-drained areas more saturated and steeper terrains better drained by gravity, with a large amount of variability in between (both in soil type and amount of soil moisture). The effects of this variability, together with the strength of the wind and other local factors, would have also produced large variability in the degree to which the ground saturation was a factor in tree downing. Clearly, more studies would be needed in order to formally show this. The discussion in Long's book, however, is valuable in more than one way. First, it adds nuance to our understanding of the effects of the Hurricane and the complexity of how these effects interact with each other to produce the damage observed, which simple one-sentence descriptions cannot properly convey. More importantly, it is also a clear example that challenging long-held assumptions in the face of conflicting evidence can lead to deeper understanding, an important process in the advance of science.

Shipwrecked in Providence

The photo used on the cover of both editions of this book is held in the Boston Public Library's Leslie Jones collection. It is captioned as a "Sunken ferry in Providence, R.I., Hurricane of 38." The collection has several other pictures of the same scene, some of which identify the ship: "Steamer Monhegan sinks at pier in Providence." A simple online search for the Monhegan easily reveals more photos of the wrecked ship, some showing its name on the hull. They also clearly show the steamer was much larger than it appears in the almost-head-on shot used on the cover. The photo was likely taken looking northwest on Peck Street, near the intersection with Dyer Street. The prominent building in the background, at 111 Westminster Street, is commonly known as the Superman Building (due to its façade resembling the Daily Planet building in the Superman comic books) and it is the tallest in the state of Rhode Island. An ordinary street is now where the ship wrecked, but during the time of the Hurricane there were a series of piers lining the west side of the Providence River (which connects with the Narragansett Bay to the south) north of the Point Street Bridge. The Monhegan became wedged between two of these piers, presumably after being wildly whipped and pushed by the higher water levels and violent surf that came with the Hurricane. The WreckHunter.net website reports it was a 128-foot passenger steamship built in 1903, and after the storm it was towed and abandoned just offshore of Prudence Island in the Narragansett Bay (approximately 150 yards north of the Sandy Point lighthouse, and exactly at 41°36'23"N, 71°18'11"W). The ship remains are under five feet of water and mostly buried in sand.

Hurricane Chairs

Chapter 9's discussion of the legacy of the storm briefly mentions furniture made with wood fallen during the Hurricane. Specifically, the furniture consists of small Windsor-style chairs made by Nichols & Stone, a Massachusetts-based company that can be traced back to at least 1752 as the Nichols Brothers Chair Manufactory. All of them are extraordinarily similar in appearance, except some are wood colored, while others are painted black, and some of the black ones have a beautiful hand-stenciled decoration. They are branded under the seat with large letters: "Pine Souvenir N.E. Hurricane Sept 21 1938" above a chair-shaped logo that includes the letters N and S (for Nichols & Stone) as part of the design.

Relief Disks

The Official Relief Relics, disks of Hurricane-fallen wood with an attached commemorative brass plate that are described and pictured in Chapter 9, hold an unexpectedly exquisite piece of historical coincidence. Upon closer inspection of the design stamped on the plate, one can discern the seal of the city of Hartford, which among other visuals, shows a deer crossing a shallow river, a "hart* ford." Under the shield containing that image, there is a ribbon with the words *post nubila phoebus*, which is the city's motto, and translates from the Latin to "after clouds sun," or if you wish, "after the clouds, comes the sun." Phoebus is an alternate name for Apollo, the Greek and Roman god of the Sun (among other titles), or at least a reference to his chief epithet of bright or brilliant. This amazingly appropriate sentiment of hope and recovery happens to be stamped (in very small lettering) on these souvenirs commemorating the Hurricane relief efforts. We have no indication one way or another, however, that the appropriate use of the phrase was intended or coincidental, as it did not become the Hartford motto as a result of the Hurricane. The motto came with the seal (designed by a local historian, Isaac William Stuart, who wrote articles in the *Hartford Courant* under the pseudonym Scaeva), which was adopted in 1852 to replace the original city seal.†,‡ Its inclusion at that time was in reference to a figurative rising from calamity, of which the city, having evolved from a colonial settlement founded in 1637, had experienced plenty.

---

* Hart is an archaic word for "stag." It was specifically used in medieval times to describe a deer stag more than five years old.

† A detailed description of the seal designed by Stuart appears in the collection of his *Hartford Courant* articles edited by W. M. B. Hartley into a book titled *Hartford in the Olden Time: Its First Thirty Years by Scaeva*, published in 1853. The article is titled "Its Name. A Coat of Arms."

‡ Information about the adoption of the Hartford seal was obtained thanks to Plymouth State University History professor Rebecca Noel.

As we make our way toward the storm's centennial, as new documents and information surface, new practices and definitions continue being developed to improve storm preparedness and response, and new storms form and cause new devastation, the Great New England Hurricane holds its place in history—it is still the one to which all hurricanes in the region are compared.

NOTES

Chapter 1

1. More information about the pre–World War II political situation can be found in Goudsouzian (2004), Chapter 6: "Aftermath."

2. A Google News search on Hitler or Chamberlain for 1938 produces many hits.

3. The report was sent as a letter from Acting Weather Bureau Chief Charles C. Clark to the Secretary of Agriculture, Henry A. Wallace (available at the National Archives at College Park textual records, Record Group 16, Item 7: Department of Agriculture General Correspondence), and transcribed here as an appendix.

4. The *New York Times*, September 22, 1938, front page.

5. See, e.g., Colton (1939).

6. See Tropical Cyclone FAQ, Subject H5, "What are the early warning signs of an approaching tropical cyclone?" on the Atlantic Oceanographic & Meteorological Laboratory (AOML) website.

7. See, e.g., the definition of "hurricane" in American Meteorological Society's *Glossary of Meteorology*.

8. See, e.g., "Hurricanes" entry in *Encyclopedia of Latin American History and Culture* (Tenenbaum 1996).

9. See, e.g., Emanuel (2005), *Divine Wind*, Chapter 3: "Huracán."

10. For more early variations of the word, see definition of "hurricane" in *Oxford English Dictionary*.

11. See the definitions of various tropical systems in National Hurricane Center Glossary.

12. Although the definition of hurricane specifies 74 mph, NHC advisories only specify wind speeds to the nearest 5 mph or knots (the lowest hurricane winds reported are 65 kt or 75 mph). The same applies to all other types of Atlantic tropical cyclones.

13. Most regions use a 10-minute average to define "sustained winds" and the various tropical cyclone classifications. The exceptions are the United States, which uses a 1-minute average, and the Australian region, which uses the maximum wind gust (3–5-second average). See, e.g., the COMET Program's online textbook *Introduction to Tropical Meteorology* (Lang and Evans 2011).

14. See, e.g., the definition of "tropical cyclone" in American Meteorological Society's *Glossary of Meteorology*.

15. The word *cyclone* was introduced in 1848 by Henry Piddington from the Greek *κύκλος* (circle) or *κυκλῶν* (moving in a circle, whirling round). See "cyclone" entry in *Oxford English Dictionary*.

16. Although not as devastating as the 1938 Hurricane, several other Atlantic tropical cyclones have affected both the Caribbean and New England, as can be found in official records starting in 1851. Irene (2011) was one of these storms.

17. See, e.g., Zelinski and Keim (2003), Chapter 18: "Hurricanes."

18. See also a detailed study of New England Hurricanes by the Blue Hill Observatory: "Climatology of Tropical Cyclones in New England and Their Impact at the Blue Hill Observatory, 1851–2009" (Iacono 2009).

19. For more information about the formation, evolution, structure, etc., of extratropical cyclones, see, e.g., Rauber et. al (any edition) *Severe and Hazardous Weather*, Chapter 10: "Extratropical Cyclones Forming East of the Rocky Mountains," and Chapter 11: "Extratropical Cyclones Forming along the East and Gulf Coasts" as well as their supporting chapters.

20. See Redfield (1831), *American Journal of Science and Arts*: article "Remarks on the Prevailing Storms of the Atlantic coast, of the North American States."

21. The observations on the nature of the southeast storms coming from the West Indies led to the development of the *Laws of Storms*, a set of laws (intended in the same fashion as the laws of motion or the laws of gravitation) describing the observed behavior of circular storms, and the subject of heated debate, the so-called *American Storm Controversy*. For more information on the American Storm Controversy, see Fleming (1990), Chapter 2: "The American Storm Controversy, 1834–1843," and Mooney (2007), Chapter 1: "Chimneys and Whirlpools."

22. On the outskirts of World War I, the term "front," meaning line of battle, was introduced to meteorological usage in 1921 by Jacob Bjerknes and Halbor Solberg. See "front" entry in *Oxford English Dictionary*.

23. A frontal boundary is defined as a cold front when cold air is advancing and replacing the warmer air.

24. A difference in density between two adjacent air masses of different temperatures implies potential energy because the air will tend to move from the area

of high density (with more molecules) toward the area with less density (with less molecules), hence acting as a force that pushes the air to move; in other words, a potential source of energy.

25. See the definition of "subtropical cyclone" in the NHC Glossary.

26. The study was done by Pierce (1939).

27. See e.g., Landsea et al. (2008).

28. The Navy and Army phonetic alphabet was used in 1950–1952. In 1952, a new international phonetic alphabet was adopted by the military, but the Weather Bureau considered it unsuitable for naming hurricanes and decided to follow the practice already in place for typhoons in the Pacific, of using female names. (See the March 1952 and September 1953 *Weather Bureau Topics*). The first list of names included Alice, Barbara, Carol, Dolly, Edna, Florence, Gail, Hazel, Irene, Jill, Katherine, Lucy, Mable, Norma, Orpha, Patsy, Queen, Rachel, Susie, Tina, Una, Vicky, and Wallis.

29. From *Historia de los Huracanes en Puerto Rico* (Miner Solá 1995).

30. See the National Hurricane Center website for a list of names to be used during the next six years as well as lists of names to be used in other regions around the world.

31. Efforts to depict and collect the tracks of all known storms in the Atlantic had been made earlier. For example, Charles L. Mitchell published various versions (e.g., 1924, 1932) of the *West Indian Hurricanes and Other Tropical Cyclones of the North Atlantic Ocean*, which compiled the tracks of all known storms in the tropical Atlantic starting in 1887.

32. The Atlantic hurricane season officially starts on June 1 and ends on November 30.

33. Detailed information about the original HURDAT (much of which also applies to HURDAT2) can be found in Landsea et al. (2004).

34. Detailed information on how to decode the original HURDAT can be found in NOAA Technical Memorandum NWS NHC 22, *A Tropical Cyclone Data Tape for the North Atlantic Basin, 1886–1983: Contents, Limitations and Uses* (Jarvinen et al. 1984).

35. For more information about HURDAT2, see AOML and NHC websites.

36. Simpson and Saffir (1974) describes the original scale.

37. See a brief discussion of factors affecting storm surge and central pressure on the NHC web page for the updated Saffir–Simpson Hurricane Wind Scale.

38. All new information about the recently redefined Saffir–Simpson Hurricane Wind Scale can be found on the NHC website.

39. Water height includes both astronomical tides and storm surge. Mean low water is basically the average low tide for a specific location.

40. See specific examples in the discussion of the updated Saffir–Simpson Hurricane Wind Scale on the NHC website.

41. See the NHC website for information about how the agency plans to deal with storm surge information in the near future.

42. See, e.g., the most updated NOAA Technical Memorandum, "The Deadliest, Costliest, and Most Intense United States Tropical Cyclones" (as of June 2013, Blake and Gibney [2011]), available on the NHC website.

43. See, e.g., the numbers in Emanuel (2005).

44. The 95% estimate is from Pierce (1939).

45. The conversion from 1938 to 2012 dollars based on the Consumer Price Index, a measure of changes in the price of goods and services that can be used to determine inflation. The CPI Inflation Calculator from the Bureau of Labor Statistics can be used to do this.

46. See, e.g., Pielke et al. (2008).

47. The $100 billion estimate based on Risk Management Solutions, Inc. (RMS 2008), report: *The 1938 Great New England Hurricane: Looking to the Past to Understand Today's Risk.*

48. The storm report for Hurricane Katrina (Knabb et al. 2006) as well as the NOAA Technical Memorandum NWS TPC-6 from the National Hurricane Center (Blake and Gibney 2011) estimate the cost at $81 billion in 2005 dollars (which would translate into $90 billion in 2010 dollars). Other estimates, including one shown on the NOAA website, put the cost up to $125 billion ($140 billion 2010 USD). (All costs in the August 2011 update of the document are given in 2010 USD. They would correspond to $95.2 billion and $147 billion 2012 USD.)

49. According to Pielke et al. (2008), accounting for changes in wealth and population as well as inflation would make the 1926 Great Miami Hurricane the costliest hurricane of all time.

50. Adapted from Colton (1939).

51. As discussed in Goudsouzian (2004).

52. General information about the economic effects of the storm can be found in many sources, e.g., Colton (1939), Scotti (2003), and Goudsouzian (2004).

53. As discussed, e.g., in Burns (2005).

54. Based on information in Scotti (2003), Chapter 21: "The Last of the Old New England Summers."

55. See, e.g., Whitnah (1961), Chapter 8: "The Modernization of General Services, 1913–41," and Harper (2008), Chapter 1: "A Stagnant Atmosphere: The Weather Services Before World War II."

56. From Gregg (1934, 1938) "Report of the Chief of the Weather Bureau," available in digital format from the NOAA Central Library.

57. The Training program was announced in June 1938 *Weather Bureau Topics and Personnel* in response to the Civil Aeronautics Act of 1938. More information on the training program, including requests for applications and announcements of those awarded in 1939 and 1940, *Topics and Personnel* issues. All available in digital format from the NOAA Central Library

58. The headquarters office of the Weather Bureau located in Washington, D.C., was known as the Central Office.

59. See information on how the Air Mass and Forecasting sections in the Washington, D.C., Office used to interact (before the Hurricane) in the 1938 Report of the Chief of the Weather Bureau (Gregg 1938).

60. According to Dunn and Gentry (1986).

61. According to Scotti (2003), Chapter 20: "The Reckoning."

62. The Bureau Chief's death and subsequent appointments are mentioned and described in the Gregg (1938) and Reicheldelfer (1939) "Report of the Chief of the Weather Bureau" and various 1938 *Weather Bureau Topics and Personnel* issues.

63. A report requested by Secretary of Agriculture Henry Wallace was sent in a letter by Charles C. Clark about two weeks after the storm, on October 3, 1938.

64. Francis Wilton Reicheldelfer was the chief of the Weather Bureau from December 1938 until his retirement in 1963. See, e.g., the NOAA Central Library web page on the Report of the Chief of the Weather Bureau for a list of Bureau Chiefs and their terms.

65. See, e.g., the "NOAA Legacy Time Line, 1900–1969" on the NOAA History website.

Chapter 2

1. From Fitzpatrick (2006), Chapter 4: "Chronology."

2. The NHC is a component of the National Centers for Environmental Prediction (NCEP) under the umbrella of the NWS (which is itself a component of NOAA, under the U.S. Department of Commerce). It is located at Florida International University in Miami, Florida, and colocated with the National Weather Service Miami–South Florida Weather Forecast Office.

3. More detailed information within the History web pages of the National Weather Service Forecast Office Miami, Florida, and the National Oceanic and Atmospheric Administration's Hurricane Research Division on their respective websites.

4. The current record for costliest hurricane season in the United States is 2005. See, e.g., the "State of the Climate, Hurricanes & Tropical Storms," 2005 annual report on the National Climatic Data Center website.

5. More information about how satellites are used to observe and study hurricanes in the COMET Program's online textbook *Introduction to Tropical Meteorology* (Lang and Evans 2011), Chapter 10: "Tropical Cyclones," section 10.4.5: "Estimation of Tropical Cyclone Intensity by Remote Sensing."

6. This and much more information about the history of hurricane hunters and what they do today, in Emanuel (2005), Chapter 25: "The Hunters," and Fitzpatrick (2006), Chapter 4: "Chronology."

7. See Tropical Cyclone FAQ, Subject H2, "Who are the Hurricane Hunters and What are they looking For?" in the Atlantic Oceanographic & Meteorological Laboratory (AOML) website.

8. All aircraft currently used for hurricane reconnaissance are described in the NOAA and Air Force hurricane hunters' respective websites.

9. The radio calling program, which commenced during the 1933 hurricane season, was announced in the May 1933 issue of the *Weather Bureau Topics and Personnel* circular.

10. Wind–pressure relationships are also used in the reanalysis efforts, which take advantage of improved regional relationships that have been developed over the years. See Landsea et al. (2004).

11. If a wind-direction observation is believed to be from a position away from the center of a hurricane, a 20° angle to the tangent direction of the circle centered on the storm is assumed, allowing a simple estimation of the general wind direction outside the center of the storm. If more than one such observation is available, it is possible to triangulate the position of said center.

12. See the discussion about "Why did the Weather Bureau not know that the storm would move so rapidly and with such great intensity over Long Island and New England" as part of the Weather Bureau Statement sent to the Department of Agriculture transcribed here as an appendix.

13. For example, the same Weather Bureau Statement described in the previous note contains a discussion about conflicting data from two vessels, which misled forecasters to think that the center of the storm was farther east than it really was.

14. For more information about the history of the usage of kites and pilot balloons, see, e.g., Fitzpatrick (2006), Chapter 4: "Chronology."

15. Fitzpatrick (2006).

16. Announced to Weather Bureau employees in the November 1938 *Weather Bureau Topics and Personnel* under the heading "New Designation for Radiometeograph Observations." Available in the NOAA Central Library Digital Documents and Maps Collections.

17. Fitzpatrick (2006).

18. Detailed information on all aspects of the Hurricane Reanalysis Project can be found in Landsea et al. (2004).

19. The completion of the reanalysis of the 1936–1940 hurricane seasons was announced in December 2012. The most updated information about the progress of the reanalysis project can be found in the project's web page within the Atlantic Oceanographic and Meteorological Laboratory (AOML)'s website.

Chapter 3

1. More than 90% of the hurricanes that reached maximum category 5 intensity on the Saffir–Simpson Scale from 1972 to 2001 were fully or partially associated with an African easterly wave. (Estimated using HURDAT and Avilés [2004], Appendix B, Table B.1.)

2. Table 1 of Avila et al. (2000) shows an average of 61 African waves per year and a range of 49 to 76 for 1972–1997.

3. In meteorology, the practice of naming direction (wind direction and direction of motion) is opposite to the practice normally followed in other sciences. For example, in physics, a vector pointing toward the east would never be named using a "west" qualifier, while in meteorology, a wind represented by the same vector would be named a "west" or "westerly" wind. (See, e.g., the vector discussion in any edition of Halliday et al. *Fundamentals of Physics* [textbook].)

4. Alternatively, they are also commonly named tropical easterly waves, African easterly waves, or simply tropical waves or African waves.

5. Temperature contrasts over Africa create a midtropospheric air current, known as the African Easterly Jet. The formation of easterly waves can be understood as shedding vortices resulting from the jet's instabilities. See, e.g., Avilés (2004) or Fitzpatrick (2006).

6. Troughs in the middle latitudes, as expected, normally look like a dip in the pressure contours (isobars), or equivalently, a dip in the height contours in a constant pressure map. In the tropics, because lower pressures are toward the equator and higher pressures toward the subtropics, the isobars must take the inverted shape such that pressure increases outward from the trough axis.

7. Day-to-day pressure variations are very small in the tropics, especially in comparison with those in the middle latitudes.

8. Herbert Riehl is informally recognized as the founder of modern tropical meteorology as its own field (Mooney 2007). He is also the author of the first textbook on tropical meteorology, aptly titled *Tropical Meteorology* (Riehl 1954). Mooney (2007) presents a fascinating portrait of the various scientists behind the major advancements in tropical meteorology from the 1800s to the present day, including Riehl, his contemporary Charney, and their pupils and current big names such as Gray and Emanuel.

9. "Waves in the easterlies" are described by Riehl (1945, 1954). The model is described in detail in Avilés (2004) as well as tested/updated for other areas besides the Caribbean. Avilés (2004) also contains a detailed description of the easterly wave life cycle and corresponding energetics.

10. Using Avila et al. (2000) Table 1 and Avilés (2004) Appendix C, Table C.1, an average of only 18% of easterly waves form a tropical cyclone at some point (18% for a system that reached at least tropical depression intensity, 10% for tropical storm intensity, and 6% for hurricanes).

11. An average of 67% of all Atlantic hurricanes from 1972 to 2001 formed from easterly waves. The percentage ranged from as low as 25% in 1972 to 93% in 1989 (Avilés 2004).

12. Dunn (1940) was the first to recognize the disturbances later named waves in the easterlies by Riehl (1945).

13. A talk presented by Charles Brooks, the director of the Blue Hill Observatory, to the American Geographical Society in November 1938 and published as Brooks (1939) did not mention this disturbance, but an updated version of Brooks's paper

appearing in Minsinger (1988) refers to a "weak low at Bilma oasis in the south-central Sahara" as the originating disturbance. The account is accompanied by a French-language reference to Hurbert (1939), who first reported the observations.

14. The energy of easterly waves over Africa comes from a combination of both direct energy transfer from the mean flow of the African Easterly Jet (known as a barotropic energy transfer) and the indirect energy transfer from available potential energy from the temperature contrasts between the hot Sahara and the cooler grasslands and ocean to the south (known as a baroclinic transfer). See Avilés (2004) for more details.

15. One way to visualize the reason for the opposite direction of the Coriolis deflection in both hemispheres is that, looking down from the North Pole, relative to its axis, the direction of Earth's rotation is counterclockwise in the Northern Hemisphere and, looking down from the South Pole, it is clockwise in the Southern Hemisphere.

16. The deflection of moving objects as viewed inside a rotating frame of reference (such as the Earth system) first appeared as a mathematical expression in an 1835 paper by French scientist Gaspard-Gustave Coriolis, but it had also been described previously by others: P. S. Laplace in 1778, G. B. Riccioli in 1651, and F. M. Grimaldi in 1651. See *Physics Today* correspondence note by Christopher Graney (2011), "Coriolis Effect, Two Centuries before Coriolis."

17. Air slows down as it moves through the trough's axis, which for an easterly flow causes what is called "speed convergence" (in similar fashion to a traffic jam as the air is forced to slow down) to the east of the axis.

18. Geographical and historical information on Cape Verde can be found on the U.S. Department of State website, in the Bureau of Public Affairs Electronic Information and Publications.

19. See, e.g., the average conditions for Praia, the capital of Cape Verde, on the BBC Weather website.

20. See Lobban (1995) for a detailed account of the history of Cape Verde.

21. Individual tropical cyclone reports for Atlantic storms can be found in the NHC online data archives.

22. Cape Verde hurricanes are briefly introduced in the NOAA/AOML Tropical Cyclone *FAQ*, Subject A2, "What is a 'Cape Verde' hurricane?" in the AOML website.

23. At 00UTC September 15 (HURDAT2 records).

24. A very small amount of information on the S.S. *Alegrete* can be found from two online sources: wrecksite.eu article, "SS *Alegrete* (I) (+1942)," and the Portuguese-language Wikipedia article, "Alegrete (navio)."

25. Tannehill (1938).

26. One hectopascal equals one millibar, which until recently was the preferred unit used by meteorologists. For the conversion to inches, see unitconversion.org and use a conversion from "inch mercury (60°F) [inHg]" to "millibars [mbar]."

27. For a tropical cyclone to produce the wind speeds necessary to be classified as a major hurricane, it must have a very low central pressure (as described in Chap-

ter 4, differences in pressure "push" the air and the larger the pressure difference, the faster the wind). The actual relationship between central pressure and tropical cyclone intensity varies from storm to storm, but it can be estimated by using wind–pressure relationships determined from statistical analysis of observed storms. Using the equations in Landsea et al. (2004) and Brown et al. (2006), a major hurricane with 100 kt maximum winds would correspond approximately to a central pressure of 960 hPa, so any pressure lower than that would suggest a major hurricane. (The lowest sea level pressure ever observed was 870 hPa during the 1979 Super Typhoon Tip in the western Pacific.)

28. The wind directions would be parallel to the isobars or lines of constant pressure. This would be correct if the ship was traveling in the eyewall. If, on the other hand, the path of the ship was somewhat more peripheral and away from the center, the wind direction would make an angle of about 20° with the isobars (Landsea et al. 2004). The conclusion of the possible path of the ship would not be too different in this case, only farther away from the center, but still skirting the northern side of the storm.

29. See Fact Sheet 6, "The Beaufort Scale," from the UK National Meteorological Library and Archive available on the UK Met Office website (Met Office 2010). The same fact sheet describes other early scales (from the 1600s and 1700s) used to estimate wind speed.

30. From Ridgway (1897) and Elligers (1903) MWR storm reports.

31. Hunter (1938) MWR report.

32. See, e.g., Rauber et al. (any edition), Chapter 24: "Tropical Cyclones" for more general information on the structure of hurricanes.

33. See Ludlum (1963) for more information on the Norfolk and Long Island Hurricane of 1921.

34. For more information on Redfield, see Longshore (2008), Fleming (1990), and Olmstead (1853).

35. Redfield (1939).

36. Redfield (1831).

37. "I chanced at that period to meet him for the first time on board a steamboat on the way from New York to New Haven. A stranger accosted me, and modestly asked leave to make a few inquiries respecting some observations I had recently published in the American Journal of Science on the subject of hailstorms. I was soon made sensible that the humble inquirer was himself a proficient in meteorology. In the course of the conversation, he incidentally brought out his theory of the laws of our Atlantic gales, at the same time stating the leading facts on which his conclusions were founded. This doctrine was quite new to me, but it impressed me so favorably, that I urged him to communicate it to the world through the medium of the American Journal of Science." (Olmstead 1853.)

38. Harvard scholar John Farrar (1818) observed that the Gale of 1815 appeared to be "a moveing [sic] vortex, and not the rushing forward of the great body of the

atmosphere" in his brief account appearing in the *Memoirs of the American Academy of Arts and Sciences*: "An account of the violent and destructive Storm of the end of September 1815."

39. Carvings of a face with two spiraling arms have been found on pottery and statues, suggesting that even before the Spanish colonization of the islands in the early 1500s, the natives of the Caribbean already recognized the devastating storms that they attributed to the evil Jurakan as rotating systems; several examples can be seen in Ortiz (2005).

40. For more on the American Storm Controversy, see Mooney (2007) and Fleming (1990).

41. Tannehill (1938).

42. From conversations with the author's grandfather, Don Alberto Avilés, during the 1970s and 1980s.

43. Fitzpatrick (2006).

44. Allen (1939).

Chapter 4

1. See, e.g., *St. Petersburg Times*, (Dec 14, 1941), "Wind and Weather Forecast Banned From Coastal Areas."

2. A summary of the changes in the Weather Bureau services because of the war can be found in the *New York Times*, (Dec 16, 1941), "Weather Bureau Curbs Service to Keep Data From Enemy's Use."

3. The National Weather Service Instruction 10-604, June 15, 2011, revised various tropical cyclone–related definitions, including hurricane watches and warnings. Additional revisions will likely occur in the future. The most current "Tropical Cyclone Definitions" are always contained in the most current version of Instruction 10-604, available on the NWS Directives website.

4. See, e.g., *The Lewiston Daily Sun*, (August 28, 1943), "'Hurricane Alerts' Will be Radioed."

5. The *Milwaukee Sentinel* article, "Hurricane Watch! U.S. Set For Danger as Season Nears," (June 28, 1956), stated that "the Weather Bureau found many persons were using the alert as an actual warning." The purpose was to "warn coastal residents that they should keep abreast of the storm's movement and watch for a hurricane warning that may be posted later."

6. The *Palm Beach Post* article, "Weather Bureau, to Drop 'Alert' in Storm Notices," (December 31, 1995), stated that "many people mistook the word *alert* as synonymous with *warning*, whereas the bureau intended it only as a preliminary notice" and "This change in terminology will also reduce misrepresentations which come from different usages of the term alert in storms advices and in civil defense and military practices."

7. "Whole gale" corresponds to 55–63 mph, force 10 in the Beaufort Scale.

8. More details about the reorganization of the Hurricane Warning Service can be found in the May 1935 *Weather Bureau Topics and Personnel.*

9. The most updated version, HURDAT2 can be found in the National Hurricane Center online data archives.

10. For more information on the Labor Day Hurricane of 1935, see, e.g., Emanuel (2005).

11. Roosevelt's New Deal programs were the subject of much political debate, especially during the 1940 presidential campaign (as evidenced in a collection of 1938–1939 public opinion surveys published by the American Institute of Public Opinion).

12. "National Disaster Staff members of Red Cross Headquarters in Washington had been notified and were already moving toward Florida, prepared to integrate their personnel with that of the local Red Cross Chapters." (Red Cross 1939.)

13. "Advisories Issued in Connection with the Tropical Hurricane of September 17–21, 1938." (C.C. Clark to H.A. Wallace, October 3, 1938, included here as an appendix.)

14. This regulation went into effect on July 1, 1935, and can be found in the May 1935 *Weather Bureau Topics and Personnel.*

15. See, e.g., Scotti (2003).

16. Norton's advisories and warnings were heard by at least 425 public outlets during the 1950s (July 1951 *Weather Bureau Topics*).

17. Most sources report that Norton died of a stroke (e.g., Fitzpatrick 2006, though incorrectly naming Dunn instead of Norton in the chronology as the one that died; October/November 1954 *Weather Bureau Topics*; and at least one newspaper article from Daytona Beach, Florida). Longshore (2008) says instead that he died of a heart attack.

18. The *Weather Bureau Topics and Personnel* (which can be found in the digital documents collection of the NOAA Central Library) was a collection of internal communications covering a wide-ranging scope and published from 1915 to 1965. The title of the publication changed throughout the years to *Weather Bureau Topics* and finally simply to *Topics.*

19. July 1954 ("Twenty Years of Hurricane Forecasting") and November 1954 ("Grady Norton") *Weather Bureau Topics* articles.

20. Dunn published his observations in a 1940 *Monthly Weather Review* article.

21. See "History of the National Weather Service Forecast Office Miami, Florida" on their website.

22. For more information about Gordon Dunn, see his biographical sketch in Fitzpatrick (2006).

23. See, e.g., the Tropical Cyclone FAQ, Subject A15, "How do tropical cyclones form?" on the AOML website.

24. This fact had been informally observed by Riehl, who stated that "condensation energy alone cannot create intense tropical storms" (Mooney 2007), and more recently Mrowiec (2011) was able to simulate a completely dry hurricane (no atmospheric water vapor) as long as the surface heat fluxes were large enough.

25. Riehl (1950) was the first to describe a hurricane as a "heat engine." Emanuel (1986) proposed a model of the hurricane as a Carnot heat engine and devised a maximum potential intensity calculation (1995, and revised in Bister and Emanuel 1998).

26. The 80°F SST threshold is the temperature above which thunderstorm formation can be supported in the tropics given current normal environmental conditions. See Tropical Cyclone FAQ, Subject A16, "Why do tropical cyclones require at least 80°F ocean temperatures to form?" on the AOML website for more information.

27. Fitzpatrick (2006).

28. The seasonal hurricane forecasts, originally done by the now-retired Bill Gray, are currently in the hands of the Tropical Meteorology Project at Colorado State University, led by Phil Klotzback. All current and past forecasts, as well as the various related publications, can be found on their website.

29. Gray (1968).

30. Technically, these storms are designated as subtropical storms and can later transition into full tropical storms (or hurricanes).

31. See the National Weather Service Online School for Weather–JetStream website, in the "Tropical Cyclones Formation Regions" topic for a listing and map of the different regions around the world where tropical cyclones form as well as a description of their corresponding hurricane season.

32. Fitzpatrick (2006).

33. See, e.g., Rauber et al. (any edition) for more on the depth of the warm layer.

34. Fitzpatrick (2006).

35. As an example, see *COMET Tropical Meteorology* (2nd ed.), Figure 8-17, for average July and December wind shear from 1958 to 2002 (Lang and Evans 2011).

36. It is common to compare winds at a pressure level of 200 hPa (close to the tropopause) and 850 hPa (about a mile above the surface). The Cooperative Institute for Meteorological Satellite Studies (CIMMS) at the University of Wisconsin–Madison Tropical Cyclone Team uses averages of winds in upper levels (150, 200, 250, 300, and 350 hPa) and lower levels (700, 775, 850, and 925 hPa) to compute and map wind shear. The maps can be found on their website.

37. See, e.g., the background information on hurricanes in Fitzpatrick (2006).

38. See, e.g., the Tropical Cyclone FAQ, Subject A17, "What is the Saharan Air Layer (SAL)? How does it affect tropical cyclones?" on the AOML website.

39. See *COMET Tropical Meteorology* (2nd ed.), Figure 8.56, for an illustration of the cyclonic circulation developed south of the SAL (Lang and Evans 2011).

40. In classical physics, the centripetal force points toward the center of rotation and the centrifugal force, pointing outward, is considered "apparent" or "imaginary." In atmospheric science, we are concerned with the forces that the air "feels," hence, we use the centrifugal force.

41. The balance between the pressure gradient, Coriolis, and centrifugal forces is called the gradient wind.

42. The balance between the pressure gradient and centrifugal forces is called cyclostrophic, and the balance between the pressure gradient and coriolis forces is called geostrophic.

43. For more information about force balances within a hurricane, see *COMET Tropical Meteorology* (2nd ed.), Section 8.2.3.1 (Lang and Evans 2011).

44. That the horizontal Coriolis force is zero at the equator (and maximum at the poles) is best shown mathematically by calculating the components of the velocity curl, which is beyond the scope of this book.

45. The vertical component, on the other hand, is maximum at the equator and zero at the poles.

46. The record for closest formation of a tropical cyclone to the equator is currently for Typhoon Vamei (2001), which formed over the western Pacific Ocean at latitude 1.5°N. Tracking information can be found in the Hurricane/Tropical Cyclone Data web page from unisys.com.

47. As suggested by Avilés (2004).

48. For more on the general circulation of the atmosphere, see, e.g., any edition of Ahrens's *Meteorology Today.*

49. More specifically, at the surface, air flow is clockwise but with an outward component, since friction of the air with the surface affects the balance of forces.

50. Besides the steering by the mean currents, hurricanes can affect their own motion by modifying their surrounding environment. There is also a weak poleward and westward drift in their motion, known as the beta drift, due to the earth's rotation. See for example Fitzpatrick (2006) for more on the topic.

51. *New York Times*, September 21, 1938: "Hurricane."

Chapter 5

1. The information about the Weather Bureau Headquarters building was obtained from a variety of sources: *Weather Bureau Topics, Fiftieth Anniversary Issue* (1941), NOAA Central Library Newsletter, Volume 1, Issue 1 (2010) and Abbe (1909) *Monthly Weather Review* chronology summary.

2. For example, the September 1936 *Weather Bureau Topics and Personnel* reported an estimated total cost of $1,201,943 for aerology services (including airway meteorological services and forecasting, upper-air soundings, and general upper-air investigations). For comparison, the general weather services and research combined (including forecasting, warning, climatological, river, forest fire, agricultural,) was $2,087,180 and the administrative costs only $138,067. This represents a total cost of $3,427,190 for the fiscal year 1936, with approximately 35.1% of the resources dedicated to aviation services.

3. In 1869 Cleveland Abbe published his first "Weather Probabilities," which is what he originally named his early forecasts. Abbe (1909), "A three-part account of the early history of meteorology in the U.S."

4. "Preliminary Report of the Special Committee on the Weather Bureau of the Science Advisory Board," November 13, 1933.

5. Gregg (1934).

6. As described by Dunn and Gentry (1986).

7. See, e.g., the definition of "synoptic" in the *Oxford English Dictionary*.

8. The methods of meteorological study usually depend on the scale being observed. Synoptic-scale phenomena (e.g., extratropical highs and lows, frontal boundaries) are of the order of about one to two thousand kilometers, roughly 500 to 1,500 miles.

9. See, e.g., Garriott (1895).

10. Ocean swells are created by distant weather systems rather than by the local winds. They are generally smoother (not white-capped) and of longer frequency.

11. See, e.g., Dole (1921) and Garriott (1900).

12. As described by Brickner (1988).

13. As described by Williams and Sheets (2001).

14. The story was pieced together from Ludlum (1963) and Williams and Sheets (2001).

15. Williams and Sheets (2001).

16. Ibid.

17. Fitzpatrick (2006).

18. Brickner (1988).

19. From observation notes by Charles F. Brooks made during the Hurricane and contained in Blue Hill Observatory private collection.

20. This and all subsequent advisories from the Weather Bureau Report to the Department of Agriculture, October 3, 1938, Clark's letter to Wallace, included here as an appendix.

21. NOAA Weather Radio can be found online or via any device that can receive the appropriate frequencies.

22. See the current Coastal Warning Display Signal light and flag signals in the NWS Marine Forecasts webpage.

23. November 1937 *Weather Bureau Topics*.

24. As described in the *Bulletin of the American Meteorological Society* (October 1995), "Necrologies."

25. Norwegan analysis techniques refer to the airmass and frontal analysis methods, which were mostly developed by Norwegian scientists in the early 20th century.

26. Even though his actual degree is not specified in any of the sources found, it could not have been specifically in meteorology or atmospheric sciences, since in 1933 the only university with such a degree was MIT (Harper 2006).

27. For a discussion on the evolution of higher education in meteorology in the United States, see Harper (2006).

28. The Library was also considered part of the professional and technical services. The Office of the Chief and a large administrative division were also part of the office (*Weather Bureau Topics, Fiftieth Anniversary Issue* [1941]).

29. A November 1, 1938, Roster of Commissioned Weather Bureau Personnel is included as part of the 1938 *Topics* file in the NOAA Central Library. The personnel listing is clearly divided by departments or divisions.

30. As reported in the November 1937 *Weather Bureau Topics*.

31. See, e.g., the information about 1938 practice forecasts appearing in the September 1937 *Weather Bureau Topics and Personnel*.

32. Report of the Chief of the Weather Bureau (Gregg 1938).

33. Ibid.

34. Neither Brickner (1988) nor Minsinger (1988) make any reference to the story.

35. The PBS *American Experience* episode "The Hurricane of '38" is identified as having first aired in 2001 (Season 13, Episode 5), but the copyright in the credits is for the year 1993. The documentary is now included as part of a 1930s collection.

36. Scotti (2003) is the most famous modern account of the storm and its surrounding events.

37. From the available sources, it is unclear if the forecaster that Pierce was replacing for the day was a "senior forecaster" or part of the "map force." Each month a different senior forecaster was in charge; it is unclear what the role of the other two senior forecasters was during that month. It seems more likely that Pierce was called in as support staff (plotting maps) in the map force.

38. Pierce (1939).

39. According to Visher (1922, from personal communications with typhoon experts Froc and Okada).

40. See, e.g., Garriott (1900), Fassig (1913), and Mitchell (1924).

41. Gregg (1938).

42. The initial publication of *Weather and Forecasting* had a biographical eulogy piece about Charles Mitchell (Dunn and Gentry [1986]). Biographical information also from *BAMS* (April 1971), "Necrology."

43. Dunn and Gentry (1986).

44. Mitchell (1924).

45. See, e.g., the discussion in the March (1939) article by Emmons about the meteorological aspects of the Hurricane.

46. William Minsinger, author of *The 1938 Hurricane: An Historical and Pictorial Summary*, recalls a conversation with Charlie Pierce in the 1980s or early 1990s, where he said that his forecast on that day just went in an envelope (Minsinger, pers. comm.).

47. As discussed by Emmons (1939).

48. Dunn and Gentry (1986).

49. This is mentioned briefly in Dunn and Gentry (1986) and verified by noting the signature of the forecaster in charge in the "Weather Conditions Last 24 Hours" portion of the Daily Weather Maps archived in the digital files within NOAA Central Library. Each senior forecaster was in charge for an entire month. There were three senior forecasters in Washington, D.C., but Mitchell's turn during the late 1930s was every other month.

50. For more on the description of the Charles L. Mitchell Award for outstanding service by a forecaster see the AMS website.

51. *BAMS* (October 1995), "Necrologies."

52. As reported by McCarthy (1969).

53. The Stewart (1939), Emmons (1939), Clowes (1939), and Burns (2005) accounts of Charles Brooks's correspondence with Mitchell all discuss the criticism.

54. The report letter from the Bureau to the Secretary of Agriculture states the following: "If plotted correctly the location of the center at 7:30 A.M. indicated that the center would pass almost directly over New York City. However, our feeling is that a warning to the effect that the center of such a storm would pass over such a populous area should not be issued until we are quite sure such will be the case. Therefore, as we considered there was still plenty of time for indicating where the storm center would pass inland . . . we awaited the noon reports."

55. Coffin (1938).

56. McCarthy (1969).

Chapter 6

1. Even though centers of high pressure are not associated with stormy weather, they are often associated with cold waves in the winter and heat waves and droughts in the summer.

2. Surface low pressure occurs when the air above weighs less, meaning that there is less air. Upper-air divergence takes air away from the column, supporting the maintenance of a surface low pressure system. The opposite is true for high pressure, which is due to more mass in the column of air above the surface. Upper-air convergence is needed to support a surface high pressure system.

3. Although today upper-air maps are drawn at constant pressure surfaces and show height contours, in the early 20th century, maps were drawn at constant height and showed isobars, or constant pressure contours.

4. Because of the centrifugal force due to the curved motion, air moves faster around ridges (where the centrifugal force adds to the effect of the pressure gradient) and slower around troughs (where it counteracts the gradient). The corresponding acceleration and deceleration produce speed divergence east of the trough and convergence west of the trough.

5. The jet stream itself is due to strong horizontal pressure changes at high levels, which are in turn caused by sharp temperature contrasts at the lower levels. The frontal boundary in which the denser, colder air to the north meets the lighter, warmer air to the south provides an increasing difference in pressure at higher altitudes and hence stronger winds.

6. As explained in Chapter 3, a trough and a ridge are elongated areas of low pressure and high pressure, respectively. They appear as dips and crests in upper-air maps.

7. The northwestward drifting tendency is known as the beta drift effect and is due to the change in the strength of Coriolis force with latitude. Hurricanes are large enough that the northward-moving air on their eastern side undergoes an increase in Coriolis force and the southward-moving air on their western side undergoes a decrease. The resulting change in force balance slightly adjusts the track northwestward. (See, e.g., Fitzgerald [2006].)

8. A clearer picture of these features can be obtained by comparing a composite of the New England hurricane hits to a composite of storms that just missed the region and noting the differences.

9. Pierce (1939).

10. For more on jet streaks and their associated convergence and divergence patterns, see, e.g., Rauber et al. (any edition).

11. It appears that the Hurricane and the trough, even being of such different size scale, might have interacted positively with each other. At the same time that the trough flow brought the Hurricane northward very fast, the Hurricane might have helped deepen the trough by enhancing the northerly flow to its west and southerly flow to its east.

12. The original report is by Tannehill (1938).

13. The haversine formula can be used to calculate the shortest distance between two points over the surface of Earth. We used the formula as available in the Movable Type Scripts website and as published by Roger Sinnott in *Sky & Telescope* magazine in 1984.

14. Speeds were calculated using latitude–longitude pairs in HURDAT/2 with the haversine formula.

15. The distance between Cape Hatteras and Long Island (as the crow flies) is approximately 400 miles. The 12Z (7 A.M.) position of 35.2°N, 73.1°W and landfall 20Z (3 PM) position of 40.7°N, 72.9°W result in a distance of 380.2 miles.

16. For more about the history of time measurement throughout the years, see, e.g., the online exhibit "A Walk Through Time: the Evolution of Time Measurement Through the Ages" by the National Institute of Standards and Technology (NIST).

17. For a detailed account of the history of Daylight Saving Time, see Prerau (2005).

18. Letter to the Editor in the *Journal of Paris*. Text available at the NIST online exhibit "Daylight Saving Time."

19. William Willett (1857–1915) wrote a pamphlet, "Waste of Daylight" (1907), that proposed advancing clocks 20 minutes on each of the four Sundays in April, and turning them back by the same amount on the four Sundays in September. The full text of the pamphlet is also included in the NIST online exhibit "Daylight Saving Time."

20. Prerau (2005).

21. For more on tornadoes, see, e.g., Rauber et al. (any edition).

22. See Tropical Cyclones FAQ, Subject L3: "What percentage of tropical cyclones spawn tornadoes?" contributed by Bill McCaul on the AOML website.

23. See Tropical Cyclones FAQ, Subject L7, "What is the largest known outbreak of TC tornadoes?" contributed by Bill McCaul on the AOML website and the references therein.

24. Gray (1919) describes a tornado within a September 1919 hurricane. The description is consistent with the first time that the phenomenon has been observed. Barbour (1924) describes a tornado/water spout within a typhoon and mentions that the phenomenon has been observed only once before.

25. See the WPC's Tropical Cyclone Rainfall Data web page.

26. From Myers and Jordan (1956) composite analysis of winds leading up to and during the landfall of the storm (Figure 10). The maps are for 10-minute winds at 30 feet. The ranges here were obtained by adjusting 12% up to 1-minute winds (as described in the Tropical Cyclone FAQ, Subject D4, within the definition of maximum sustained winds) in the AOML website.

27. For more about the Blue Hill Meteorological Observatory, see their "Brief History" web page. Most data date to 1885 or 1886, but temperature records extend to 1831.

28. Sustained winds, when referring to observed (rather than estimated), correspond to the 2-minute-averaged winds reported by automated weather stations. The difference between 1-minute- and 2-minute-averaged wind is insignificant. Therefore, no attempt is made to convert them. The rest of the world (besides the United States) uses 10-minute averages, which are significantly lower and must be converted accordingly for any comparison between storms to be significant. See the Tropical Cyclones FAQ, Subject D4: "What does 'maximum sustained wind' mean? How does it relate to gusts in tropical cyclones?" contributed by Chris Landsea on the AOML website and the references therein.

29. This corresponds to half of the 12% specified above for a 10-minute to 1-minute conversion.

30. For example, in 1938 the anemometer at the Weather Bureau Headquarters in D.C. was at a height of 85 feet, and at the airport, it was at 23 feet, as reported by Grice (2005).

31. Which can be done by using graphs and information from Franklin et al. (2003).

32. As included in the recently revised HURDAT2.

33. As defined in the American Meteorological Society *Glossary of Meteorology*.

34. See Emanuel (2005), e.g., for the equivalency between pressure drop inside the eye to sea level height increase.

35. See the Tropical Cyclones FAQ, Subject C1, on the factors that affect storm surge (contributed by the NHC Storm Surge Unit) on the AOML website.

36. Although the relationship between category, wind speed, and storm surge (and central pressure) has been removed from the Saffir–Simpson scale, tables with all of these quantities can still be easily found online and in textbooks published before 2010.

37. The NHC uses the SLOSH (Sea, Lake, and Overland Surges from Hurricanes) model, developed by the National Weather Service to estimate storm surge heights resulting from hurricanes. See the NHC's web page on the SLOSH model for more information.

38. USGS (1940).

39. Various standard tidal "datums" can be used as reference points to report water height: Mean High Water (MHW), Mean Higher High Water (MHHW), Mean Sea Level (MSL), Mean Low Water (MLW), and Mean Lower Low Water (MLLW).

40. A tidal epoch is a 19-year period over which tide observations are taken and used to obtain mean values. It corresponds to the period of time over which the gravitational effects resulting from the relative positions of Earth and the moon and sun repeat.

41. In 1999, NOAA calculated the percentage of various causes of tropical cyclone deaths in the United States from 1970 to 1999 as a way to assess modern hurricane-related dangers. Freshwater flooding was by far the largest killer with 59%. The storm surge accounted for only 1% and the surf for 11%, as calculated by Edward Rappaport of the Technical Support Branch of the NHC (then the Tropical Prediction Center). While it is still true that in modern times freshwater flooding causes hurricane-related deaths most often, the fatalities due to Hurricane Katrina's (2005) surge after the breaking of the levees in New Orleans were so numerous (1,200) that they would represent a huge percentage of any new statistics being computed (as briefly discussed in NOAA Technical Memorandum, NWS NHC-6).

42. Many articles from the time refer to the storm surge as the "storm wave." See, e.g., USGS (1940).

43. The "dangerous semicircle" and the "navigable semicircle" are defined in the American Meteorological Society's *Glossary of Meteorology*.

Chapter 7

1. Calculated with the original HURDAT, 46.7% of all tropical cyclones in the database from 1980 to 2012 were classified as extratropical toward the end of their lifetime.

2. Visher (1922) specifically talks about Froc and Okada's opinion that most typhoons undergo extratropical transition, and it was a matter of time before a study could confirm this once sufficient data was available.

3. NWSI 10-604, "Tropical Cyclone Definitions," available on the National Weather Service Directive Systems website.

4. "LO" is an option for "Status of system" included in spaces 20–21 of the database file. It is defined as a low that is not a tropical cyclone, a subtropical cyclone, or an extratropical cyclone (of any intensity). Therefore, it is not only used for remnant lows but also for low pressure areas preceding the tropical cyclone formation. (Landsea et al. [2013] documentation on the revised HURDAT2.)

5. Pierce (1939).

6. For example, Neumann et al. (1999) as discussed in Landsea et al. (2008).

7. Landsea et al. (2008).

8. USGS (1940).

9. From Massachusetts Department of Public Health, as included in USGS (1940) report (not in NCDC data).

10. Galarneau et al. (2010) first defined predecessor rain events ahead of tropical cyclones.

11. As can be seen in their respective rainfall maps, available from the Weather Prediction Center website on Tropical Cyclone Rainfall, which contains total rainfall maps for all Atlantic tropical cyclones dating to 1964, as well as other historical storms.

12. As presented in a 2012 Northeastern Storm Conference presentation by Paul A. Sisson of the NWS Burlington, VT, office on "Tropical Cyclone Irene's Devastating Flash Flooding in Vermont."

13. 1938 September normals obtained from NCDC.

14. Currently, "normal" is calculated from 30-year averages of the previous three complete decades (the current normals are for 1981–2010, e.g.) and distinguished from record-length averages, usually referred to as the "means."

15. Calculated with data from USGS (1940) Table 2.

16. Calculated using NCDC September 1938 normals.

17. From NWS Eastern Region Headquarters web page on Historical Floods in the Northeast.

18. USGS (1940).

19. The Pemigewasset River in central New Hampshire is a tributary of the Merrimack River, which crosses south-central New Hampshire to the Massachusetts border, draining into the Atlantic Ocean.

20. From the "Historical Crests for the Pemigewasset River at Plymouth," available from the National Weather Service's Advanced Hydrologic Prediction Center's web page.

21. USGS (1940).

22. USGS gaging station No. 01076500.

23. Historical river flow data can be obtained from the USGS National Water Information System web page.

24. Blowdown estimates from Forest Service (1943).

25. Ibid.

26. The volume of a cylinder (our tree) is $\pi r^2 h$, where r is the log's radius and h is the height. Using inches, this equation produces a total of 327.2 board feet per tree.

27. *Forest Service Timber Management Field* (2008), Table 1-2, shows the number of board feet for 16-foot logs of a certain chest height diameter.

28. Using three 16-foot logs (to simulate 50 feet high) and chest height diameter of 10 inches, the estimated number of board feet per tree would be 1,485.

29. U.S. Forest Service (1943).

30. Foster and Boose (1992).

31. Cooper-Ellis et al. (1999).

32. As reported by New Hampshire Disaster Emergency Committee (1938).

33. Foster (1992).

34. Reported in some sources as 90 feet and in others as 175 feet tall.

35. Information provided by the New Hampshire Department of Transportation web page on New Hampshire Covered Bridges.

36. Bristol Enterprise editorial, as reported in New Hampshire Disaster Emergency Committee (1938).

37. "The Hurricane and Flood of 1938 by Girls and Boys Connecticut Supervisory District #7."

38. Speare (1963).

39. Poole (1946).

40. Speare (1963).

Chapter 8

1. It was classified as an extratropically transitioned tropical storm in the original HURDAT (Tropical Storm #7). It was removed from the database and not included in the new HURDAT2, after the completion of the reanalysis of the 1938 hurricane season.

2. Brown (1939).

3. The height to which debris was deposited is given by Nichols and Marston (1939).

4. Initial estimates by the Red Cross (1939) include a breakdown of casualties by state in their serviced area: 207 RI; 117 MA; 97 CT; 60 NY; 12 NH; 1 VT; and 0 NJ.

5. Clowes (1939).

6. The lowest estimate found is 494 from the Red Cross (1939) and the highest is 682 from the WPA (1938), the latter often being considered as the official number of deaths from the storm.

7. Account by Ms. Carol Sandford, originally from the school's Christmas magazine *The Crystal*, as reproduced in Clowes (1939).

8. Clowes (1939).

9. WPA (1938).

10. Ibid.

11. A short circuit occurs when two portions of an electrical circuit that are not meant to connect become connected. In this case the flooding water provided the unintended connection.

12. WPA (1938).

13. As the grain particles come close to each other, electrons can jump from one to the other causing an excess of electrons (or negative charge) in some and a deficit (or positive charge) in others. A spark can then occur between the negative and positive charges.

14. From *Providence Journal* behind-the-scenes brochure, "The Story Within" picture caption: "One Newsman writes his story while others hold lanterns during lightless hours of emergency."

15. See, e.g., historybuff.com for more information about the history of typesetting technology.

16. *Providence Journal*, "The Story Within."

17. Zielinski and Keim (2003), "New England Weather, New England Climate."

18. Brown (1939).

19. Since the last catastrophic event in the area, the Great Gale of 1815.

20. Brown (1939).

21. Nichols and Marston (1939).

22. Clowes (1939).

23. WPA (1938).

24. Forest Service (1943).

25. Ibid.

26. Ibid.

27. The information is as found in Smith (2010), which used Forest Service (1943) and personal accounts as sources.

28. Forest Service (1943).

29. Coffin (1942).

30. Forest Service (1943).

31. The watershed of a body of water is the surrounding terrain that drains into it.

32. Foster and Boose (1995).

33. Foster et al. (1997).

34. See "About the Agency" on the FEMA website for more information about the agency's mission and history.

35. WPA (1938).

36. Ibid.

37. Ibid.

38. A congressional charter is a law passed by U.S. Congress that states the mission, authority, and activities of a group.

39. The entire National Response Framework document can be found on the FEMA website.

40. A Consumer Price Index Inflation Calculator can be found on the Bureau of Labor Statistics' Databases, Tables & Calculators web page.

41. Crooks (1939) and WPA (1938).

42. Red Cross (1939).

Chapter 9

1. For detailed information on the HURISK model, see NOAA Technical Memorandum NWS NHC 38 (Neumann 1987).

2. Ludlum (1963) and Donnelly et al. (2001).

3. Ludlum (1963).

4. Ibid.

5. Perley (1891).

6. Forest Service (1943) and Perley (1891).

7. A variety of stratigraphic and isotopic dating methods were used. Stratigraphic techniques use the fact that oldest layers are deposited first and hence are found deeper in sediment cores. They also use cross-referencing of similar characteristics with sites of known date. Isotopic methods use the ratio of isotopes (same element, but different weights—due to different number of neutrons in their nuclei) present and knowledge of their relative decay rates to determine the age of a sample.

8. Bravo et al. (1997) and Donnelly et al. (2001).

9. Besides the intrinsic uncertainties associated with stratigraphic dating techniques, according to Donnelly et al. (2001) there is also the possibility that a portion of the deposit layer attributed to the 1635 Hurricane was deposited in 1638, as the historical records (as reported by Ludlum) speak of two or three storms, at least one of which came with an extremely high tide during that year.

10. Jarvinen (2006).

11. Or even greater, given observed and expected higher sea levels during the coming decades and hence higher potential storm surges.

12. As reported by Pielke (2008).

13. NOAA Technical Memorandum NWS NHC-6 was originally published in 1997 (for storms from 1886 to 1995) and has been updated various times since. The latest update includes tropical cyclones from 1851 to 2010 and was published in August 2011.

14. With a preliminary estimated cost of $71 billion, Hurricane Sandy will most likely slide into the number 2 ranking for cost only adjusted by inflation, and number 4 or 5 for cost adjusted for inflation, population, and wealth normalization (Table 3b of NOAA Technical Memorandum NWS NHC-6, Blake and Gibney [2011]).

15. RMS (2008).

16. Pielke et al. (2008).

17. RMS (2008).

18. The costliest disaster in history is the earthquake and tsunami in Japan on March 11, 2011, which led to the failure of the Fukushima nuclear power plant. Within two weeks of the disaster, the World Bank estimated that the cost was approximately $235 billion (as reported by the *Washington Post* on March 21, 2011).

19. See the official error trends in the NHC Tropical Cyclone Forecast Verification web page.

20. According to NOAA's August 30, 2011, summary, *Irene by the Numbers*, "Because NHC did not anticipate the slow weakening between the Bahamas and North Carolina, early forecasts for the Mid-Atlantic states and New England were too high. Although subsequent intensity forecasts were lower, they never quite caught up with the actual weakening of the storm."

21. It also made it clear that the public was confused about the lack of warnings and the severity of the storm hazards. The latest revision of warning practices, done almost immediately following the 2012 hurricane season and in effect for the 2013 season, allows the continuation of warnings for post-tropical systems and should take care of avoiding this confusion in the future.

22. From personal communication with Gregory Champlin from the New Hampshire Emergency Management Office.

23. From personal communication with Stephen Long, founding editor of *Northern Woodslands* magazine and Harvard University's Harvard Forest fellow.

24. Hale (2010).

25. Plymouth and various other New Hampshire towns were established in the year 1763, which was also when the French and Indian War ended, making frontier land less dangerous and available for settlement. "Marking the Moment," an original musical commemorating Plymouth's 250th anniversary, was a presentation of the Educational Theater Collaborative, January 23–26, 2013, written by Manuel Marquez-Sterling and Trish Lindberg (also the director) and music by William Ogmundson.

26. Red Cross (1939).

REFERENCES

Abbe, C., 1899: *The Aims and Methods of Meteorological Work.* The Johns Hopkins University Press, 330 pp.

———, 1909: A Chronological Outline of the History of Meteorology in the United States of North America. *Mon. Wea. Rev.*, **37**, 87–89, 146–149, 178–180.

Ahrens, C. D., 2009: *Meteorology Today: An Introduction to Weather, Climate, and the Environment* (9th ed). Brooks/Cole, 624 pp.

Allen, G. M., 1939: Hurricane Aftermath. *The Auk*, **56**, 176–179.

American Association for Public Opinion Research, 1939: American Institute of Public Opinion Surveys, 1939–1939. *Public Opinion Quarterly*, **3**, 581–507.

American Meteorological Society, 1995: Necrologies: Charles H. Pierce 1909–1994. *Bull. Amer. Meteor. Soc.*, **76**, 1838.

———, 1971: Necrology: Charles L. Mitchell 1883-1970. *Bull. Amer. Meteor. Soc.*, **52**, 257.

American National Red Cross, 1939: New York – New England Hurricane and Floods – 1938: Official Report of Relief Operations, 106 pp.

Avila, L. A., R. J. Pasch, and J. G. Jing, 2000: Atlantic Tropical Systems of 1996 and 1997: Years of Contrasts. *Mon Wea. Rev.*, **128**, 3695–3706.

Avilés, L. B., 2004: African Easterly Waves: Evolution and Relationship to Atlantic Tropical Cyclones. Ph.D. Diss., University of Illinois, 211 pp.

Bister, M., and K. A. Emanuel, 1998: Dissipative Heating and Hurricane Intensity. *Meteor. Atmos. Phys.*, **65**, 233–240.

Blake, E. S., and E. J. Gibney, 2011: The Deadliest, Costliest, and Most Intense United States Tropical Cyclones from 1851 to 2010 (and Other Frequently Requested Hurricane Facts). NOAA Tech. Memo. NWS NHC-6, 47 pp.

Bravo, J., J. P. Donnelly, J. Dowling, and T. Webb III, 1997: Lithologic and Biostratigraphic Evidence of the 1938 Hurricane Event in New England. Preprints, *22nd Conf. on Hurricanes and Tropical Meteorology*, Ft. Collins, CO, Amer. Meteor. Soc., 395–396.

Brickner, R. K., 1988: *The Long Island Express: Tracking the Hurricane of 1938*. Hodgins Printing Company, 125 pp.

Brooks, C. F., 1938: Hurricanes into New England: Meteorology of the Storm of September 21, 1938. Address to the American Geographical Society on November 1, 1938. Published 1939 in *Geogr. Rev.*, **29**, 119–127.

———, 1938: The Hurricane of September 21, 1938: Observations at Blue Hill Observatory, Milton, Mass. Report transcript of observer notes and meteorological observations, Blue Hill Observatory private collection, 4 pp.

Brown, C. W., 1939: Hurricanes and Shore-Line Changes in Rhode Island. *Geogr. Rev.*, **XXIX**, 416–430.

Brown, D. P., J. L. Franklin, and C. W. Landsea, 2006: A Fresh Look at Tropical Cyclone Pressure-Wind Relationships Using Recent Reconnaissance-based "best track" data (1998–2005). Preprints, *27th Conf. on Hurricanes and Tropical Meteorology*, Monterey, CA, Amer. Meteor. Soc., 3B.5.

Burns, C., 2005: *The Great Hurricane: 1938*. Grove Press, 230 pp.

Clark, C. C., 1938: The Tropical Storm of September 17–21, 1938. Memorandum for the Secretary of Agriculture. October 3, 1938, 16 pp.

Clowes, E. S., 1939: *The Hurricane of 1938 on Eastern Long Island*. Hampton Press, 71 pp.

Coffin, J. E., 1938: *It Did Happen Here! An Illustrated Review of the Damage Wrought in the Western Part of the Monadnock Region by the Hurricane and Flood of 1938*. Granite State Studio and Sentinel Printing Company, 76 pp.

———, 1942: After the Big Blow-Down. *Outdoor Life*, **89**, 26–27, 56–57.

Colton, F. B., 1939: The Geography of a Hurricane. *National Geographic Magazine*, **LXXV**, 529–552.

Connecticut State Department of Education, 1939: The Hurricane and Flood of 1938 by Girls and Boys Connecticut Supervisory District #7, 87 pp.

Cooper-Ellis, S., D. R. Foster, G. Carlton, and A. Lezberg, 1999: Forest Response to Catastrophic Wind: Results from an Experimental Hurricane. *Ecology*, **80**, 2683–2696.

Dole, R. M., 1921: The Fire-Colored Sunset as a Valuable Clue to the Existence of a Tropical Storm. *Mon. Wea. Rev.*, **49**, 191.

Donnelly, J. P., S. S. Bryant, and J. Butler, et al., 2001: 700 Yr Sedimentary Record of Intense Hurricane Landfalls in Southern New England. *Geol. Soc. Amer. Bull.*, **113**, 714–727.

Dunn, G. E., 1940: Aerology in the Hurricane Warning Service. *Mon. Wea. Rev.*, **68**, 303–315.

———, and R. C. Gentry, 1986: Charles L. Mitchell: Remarkable Forecaster – Rare Friend (Forecaster's Biography). *Wea. Forecasting*, **1**, 108–110.

Emanuel, K., 2005: *Divine Wind: The History and Science of Hurricanes*. Oxford University Press, 285 pp.

Emmons, G., 1939: The Meteorological Aspects of the New England Hurricane. Hurricane Number. *The Collecting Net*, **14** (Suppl.), 2–9.

Farrar, J., 1818: An Account of the Violent and Destructive Storm of the 23rd of September 1815. *Mem. Amer. Acad. Arts Sci.*, **IV–I**, 92–97.

Fitzpatrick, P. J., 2006: *Hurricanes: A Reference Handbook* (2nd ed.), ABC CLIO, 412 pp.

Fleming, J. R., 1990: *Meteorology in America, 1800–1970*. The Johns Hopkins University Press, 264 pp.

Forest Service, 1943: Report of the U.S. Forest Service Programs Resulting from the New England Hurricane of September 21, 1938. Northeastern Timber Salvage Administration, 593 pp.

———, 2008: Timber Management Field Book, 141 pp.

Foster, D. R., 1992: Land-Use History (1730–1990) and Vegetation Dynamics in Central New England, USA. *J. Ecol.*, **80**, 753–772.

———, and E. R. Boose, 1992: Patterns of Forest Damage resulting from Catastrophic Wind in Central New England, USA. *J. Ecol.*, **80**, 79–98.

———, 1995: Hurricane Disturbance Regimes in Temperate and Tropical Forest Ecosystems. *Wind Effects of Trees, Forests and Landscapes*, M. Coutts, Ed., Cambridge University Press, 305–339.

Foster, D. R., J. D. Aber, J. M. Melillo, R. D. Bowden, and F. A. Bazzaz, 1997: Forest Response to Disturbance and Anthropogenic Stress. *BioScience*, **47**, 437–445.

Franklin, B., 1784: To the Authors of the Journal of Paris. *The Ingenious Dr. Franklin. Selected Scientific Letters*, N. G. Goodman, Ed., University of Pennsylvania Press, 17–22.

Franklin, J. L., M. L. Black, and K. Valde, 2003: GPS Dropwindsonde Wind Profiles in Hurricanes and Their Operational Implications. *Wea. Forecasting*, **18**, 32–44.

Galarneau, T. J., L. F. Bosart, and R. S. Schumacher, 2010: Predecessor Rain Events ahead of Tropical Cyclones. *Mon. Wea. Rev.*, **138**, 3272–3297.

Garriott, E. B., 1895: Tropical Storms of the Gulf of Mexico and the Atlantic Ocean in September. *Mon. Wea. Rev.*, **23**, 167–169.

Goudsouzian, A., 2004: *The Hurricane of 1938*. Commonwealth Editions, 90 pp.

Gray, R. W., 1919: A Tornado within a Hurricane Area. *Mon. Wea. Rev.*, **47**, 639.

Gray, W. M., 1968: Global View of the Origin of Tropical Disturbances and Storms. *Mon. Wea. Rev.*, **96**, 669–700.

Gregg, W. R., 1934: Progress in Development of the U.S. Weather Service in Line with the Recommendations of the Science Advisory Board. *Science*, **80**, 349–351.

———, 1934: Report of the Chief of the Weather Bureau, 7 pp. [Available online from NOAA Central Library at www.lib.noaa.gov/collections/imgdocmaps/reportofthechief.html.]

———, 1938: Report of the Chief of the Weather Bureau, 15 pp. [Available online from NOAA Central Library at www.lib.noaa.gov/collections/imgdocmaps/reportofthechief.html.]

Grice, G. K., 2005: History of Weather Observing in Washington, D.C. 1821–1950. Midwestern Regional Climate Center, Climate Database Modernization Program, NOAA NCDC, 35 pp.

Grossi, P., 2008: The 1938 Great New England Hurricane: Looking to the Past to Understand Today's Risk. Report, Risk Management Solutions, Inc., 18 pp.

Hale, J. D., 2010: *Inside New England: The Editor-in-Chief of Yankee Magazine Reveals the New England Known Only to Locals*. Bauhan Publishing LLC, 272 pp.

Halverson, J. B., and T. Rabenhorst, 2013: Hurricane Sandy: The Science and Impacts of a Superstorm. *Weatherwise*, **66**, 14–23.

Harper, K. C., 2006: Meteorology's Struggle for Professional Recognition in the USA (1900–1950). *Ann. Sci.*, **63**, 179–199.

———, 2008: *Weather by the Numbers: The Genesis of Modern Meteorology*. MIT Press, 308 pp.

Hobbs, F., and N. Stoops, 2002: Demographic Trends in the 20th Century. Census 2000 Special Reports, U.S. Census Bureau, CENSR-4, 222 pp.

Hunter, H. C., 1938: Weather on the Atlantic and Pacific Oceans. *Mon. Wea. Rev.*, **66**, 299–300.

Iacono, M. J., 2009: Climatology of Tropical Cyclones in New England and Their Impact at the Blue Hill Observatory, 1851–2009. Blue Hill Observatory Report, 13 pp.

Jarvinen, B. R., C. J. Neumann, and M. A. S. Davis, 1984: A Tropical Cyclone Data Tape for the North Atlantic Basin, 1886–1983: Contents, Limitations, and Uses. NOAA Tech. Memo. NWS NHC 22, 21 pp.

———, 2006: Storm Tides in Twelve Tropical Cyclones (including Four Intense New England Hurricanes). National Hurricane Center Report, 99 pp.

Knabb, R. D., J. R. Rhome, and D. P. Brown: Updated 2006: Hurricane Katrina. NHC Tropical Cyclone Report, 43 pp.

Laing, A., and J. L. Evans, 2011: *Introduction to Tropical Meteorology: A Comprehensive Online & Print Textbook*(2nd ed.), COMET Program, University Corporation for Atmospheric Research. [Available online at http://www.meted.ucar.edu/tropical/textbook_2nd_edition/index.htm.]

Landsea, C. W., C. Anderson, N. Charles, et al., 2004: The Atlantic Hurricane Database Reanalysis Project: Documentation for 1851–1910 Alterations and Additions to the HURDAT Database. *Hurricanes and Typhoons: Past, Present and Future*, R. J. Murnane and K. B. Kiu, Eds., Columbia University Press, 177–221.

———, M. Dickinson, and D. Strahan, 2008: Reanalysis of Ten U.S. Landfalling Hurricanes. Risk Prediction Initiative Report, 120 pp.

———, J. Franklin, and J. Beven, 2013: The Revised Atlantic Hurricane Database (HURDAT2). NHC/AOML/HRD Note. [Available online at www.nhc.noaa.gov/pastall.shtml#hurdat.]

Lobban, R. A., 1998: *Cape Verde: Crioulo Colony to Independent Nation*. Westview Press, 200 pp.

Longshore, D., 2008: *Encyclopedia of Hurricanes, Typhoons, and Cyclones, New Edition*. Checkmark Books, 468 pp.

Ludlum, D. M., 1963: *Early American Hurricanes: 1492–1870*. American Meteorological Society, 198 pp.

Mackun, P., and S. Wilson, 2011: Population Distribution and Change: 2000 to 2010. 2010 Census Briefs, U.S. Census Bureau, C2010BR-01, 11 pp.

McCarthy, J., 1969: *Hurricane!* American Heritage Press, 168 pp.

Met Office, 2010: The Beaufort Scale. National Meteorological Library and Archive Fact Sheet 6, 20 pp.

Miner Solá, E., 1995: *Historia de los Huracanes en Puerto Rico*. First Book Printing of P.R., 94 pp.

Minsinger, W. E., 1988: *The 1938 Hurricane: An Historical and Pictorial Summary*. Blue Hill Observatory, 128 pp.

Mitchell, C. L., 1924: West Indian Hurricanes and Other Tropical Cyclones of the North Atlantic Ocean. *Mon. Wea. Rev.*, **24** (Suppl.), 47 pp.

———, 1932: West Indian Hurricanes and Other Tropical Cyclones of the North Atlantic Ocean. *Mon. Wea. Rev.*, **60**, 253.

Mooney, C., 2007: *Storm World: Hurricanes, Politics, and the Battle Over Global Warming*. Harcourt, Inc., 400 pp.

Mrowiec, A. A., S. T. Garner, and O. M. Pauluis, 2011: Axisymmetric Hurricane in a Dry Atmosphere: Theoretical Framework and Numerical Experiments. *J. Atmos. Sci.*, **68**, 1607–2011.

Myers, V. A., and E. S. Jordan, 1956: Winds and Pressures over the Sea in the Hurricane of September 1938. *Mon. Wea. Rev.*, **84**, 261–270.

National Climatic Data Center, 2005: State of the Climate: Hurricanes & Tropical Storms for Annual 2005 (published online December 2005, retrieved on April 13, 2013). [Available online at http://www.ncdc.noaa.gov/sotc/tropical-cyclones/2005/13.]

National Weather Service, 2011: Tropical Cyclone Definitions. NWS Instruction NWSPD 10-604, June 15, 2011, 13 pp. [Current version available online at www.nws.noaa.gov/directives/sym/pd01006004curr.pdf.]

Neumann, C. J., 1987: The National Hurricane Center Risk Analysis Program (HURISK). NOAA Tech. Memo. NWS NHC 38, 56 pp.

New Hampshire Disaster Emergency Committee, 1938: The Flood and Gale of September 1938, 92 pp.

Nichols, R. L., and A. F. Marston, 1939: Shoreline Changes in Rhode Island Produced by Hurricane of September 21, 1938. *Bull. Geol. Soc. Amer.*, **50**, 1357–1370.

Olmsted, D., 1857: Address on the Scientific Life and Labors of William C. Redfield, First President of the American Association for the Advancement of Science. Delivered before the Association at their Annual Meeting, August 14, 1857, 28 pp.

Ortiz, F., 2005: *El Huracán: Su mitología y sus símbolos*. Fondo de Cultura Económica, 533 pp.

Perley, S., 1891: *Historic Storms of New England*. The Salem Press Publishing and Printing Company, 341 pp.

Pielke, R. A., J. Gratz, C. W. Landsea, D. Collins, M. A. Saunders, and R. Musulin, 2008: Normalized Hurricane Damage in the United States: 1900–2005. *Nat. Hazards Rev.*, **9**, 29–42.

Pierce, C. H., 1939: The Meteorological History of the New England Hurricane of Sept. 21, 1938. *Mon. Wea. Rev.*, **67**, 237–285.

Poole, E., 1946: *The Great White Hills of New Hampshire*. Doubleday, 472 pp.

Prerau, D., 2005: *Seize the Daylight: The Curious and Contentious Story of Daylight Saving Time*. Basic Books, 256 pp.

Providence Journal-Bulletin: The Hurricane of 1938: How Readers Were Supplied With News in the Midst of Disaster. The Story Within, 14–15

Raines, R. R., 1996: *Getting the Message Through: A Branch History of the U.S. Army Signal Corps*. Government Printing Office, 464 pp.

Rauber, R. M., J. E. Walsh, and D. J. Charlevoix, 2008: *Severe & Hazardous Weather: An Introduction to High Impact Meteorology* (3rd ed.), Kendall/Hunt Publishing Company, 642 pp.

Redfield A. C., 1939: The Hurricane and Others. *The Collecting Net*, **14** (Suppl.), 10–14.

Redfield. W. C., 1831: Remarks on the Prevailing Storms of the Atlantic Coast, of the North American States. *Amer. J. Sci. Arts*, **20**, 17–51.

Reichelderfer, R. W., 1939: Report of the Chief of the Weather Bureau, 21 pp. [Available online from NOAA Central Library at www.lib.noaa.gov/collections/imgdocmaps/reportofthechief.html.]

Riehl, H., 1950: A Model for Hurricane Formation. *J. Appl. Phys.*, **21**, 917–925.

———, 1954: *Tropical Meteorology*. McGraw-Hill Book Company, 392 pp.

Rolt-Wheeler, F. W., 1917: *The Boy with the U.S. Weather Men*. U.S. Service Series. Lothrop, Lee & Shepard Co., 336 pp.

Science Advisory Board, 1933: Preliminary Report of the Special Committee on the Weather Bureau of the Science Advisory Board, November 13, 1933, SAB, 16 pp.

Scotti, R. A., 2003: *Sudden Sea: The Great Hurricane of 1938*. Back Bay Books, 281 pp.

Simpson, R. H., and H. Saffir, 1974: The Hurricane Disaster-Potential Scale. *Weatherwise*, **27**, 169.

Smith S. S., 2010: *They Sawed up a Storm*. Peter E. Randall Publisher, 66 pp.

Speare, E. A., 1963: *Twenty Decades in Plymouth, New Hampshire, 1763–1963*. New England History Press, 183 pp.

Stewart, J. Q., 1939: New England Hurricane. *Harper's Magazine*, No. 1064 (January 1939), 198–204.

Tannehill, I. R., 1938: Hurricane of September 16 to 22, 1938. *Mon Wea. Rev.*, **66**, 286–288.
Tenenbaum, B. A., 1996: *Encyclopedia of Latin American History and Culture*, Vol. 3, Simon & Schuster Macmillan, 596 pp.
United States Geological Survey, 1940: Hurricane Floods of September 1938. USGS Water-Supply Paper 867, 562 pp.
Visher, S. S., 1922: Notes on Typhoons, with Charts of Normal and Aberrant Tracks. *Mon. Wea. Rev.*, **50**, 583–589.
Weather Bureau, 1933, 1935, 1936, 1937, 1938, 1939, 1940, 1941, 1951, 1952, 1953, 1954: Topics and Personnel. [Available from NOAA Central Library Digital Documents at www.lib.noaa.gov/collections/imgdocmaps/topicsandpersonnel.html.]
Whitnah, D. R., 1961: *A History of the United States Weather Bureau*. University of Illinois Press, 267 pp.
Williams, J., and B. Sheets, 2001: *Hurricane Watch: Forecasting the Deadliest Storms on Earth*. Random House, 352 pp.
Willett, W. M., 1907: The Waste of Daylight. Essay reprinted in D. de Carle, 1946: *British Time*. Crosby Lockwood & Son, Ltd., 152–157.
Works Progress Administration/Federal Writers' Project, 1938: New England Hurricane: A Factual, Pictorial Record. WPA, 220 pp.
Zielinski, G. A., and B. D. Keim, 2003: *New England Weather, New England Climate*. University Press of New England, 276 pp.

Online Resources

(all accessed June 2013)

Atlantic Oceanographic & Meteorological Laboratory—Hurricane Research Division HRD. http://www.aoml.noaa.gov/hrd/. Tropical Cyclone Frequently Asked Questions (FAQ). http://www.aoml.noaa.gov/hrd/tcfaq/tcfaqHED.html. Hurricane Reanalysis Project. http://www.aoml.noaa.gov/hrd/data_sub/re_anal.html.
Blue Hill Observatory. http://www.bluehill.org. Climate Data and Research. http://www.bluehill.org/climate/climate.html.
Bureau of Labor Statistics—CPI Inflation Calculator. http://www.bls.gov/data/inflation_calculator.htm.
Colorado State University Tropical Meteorology Project. http://hurricane.atmos.colostate.edu.
Haversine Formula—Movable Type Scripts. http://www.movable-type.co.uk/scripts/latlong.html.
Hurricane Hunters Related: 53rd Weather Reconnaissance Squadron Factsheet. http://www.403wg.afrc.af.mil/library/factsheets/factsheet.asp?id=7483. NOAA Aircraft Operations Center—Aircraft. http://www.aoc.noaa.gov/aircraft.htm.
Meteorological Glossaries: American Meteorological Society Glossary of Meteorology. http://amsglossary.allenpress.com/glossary. National Hurricane Center

Glossary of Terms. http://www.nhc.noaa.gov/aboutgloss.shtml. National Weather Service Glossary. http://w1.weather.gov/glossary/.

National Hurricane Center. http://www.nhc.noaa.gov/. Data Archive. http://www.nhc.noaa.gov/pastall.shtml. The Saffir–Simpson Hurricane Wind Scale. http://www.nhc.noaa.gov/sshws.shtml. Statement on Storm Surge Scales and Storm Surge Forecasting. http://www.nhc.noaa.gov/pdf/sshws_statement.pdf. SLOSH Model. http://www.nhc.noaa.gov/surge/ssurge_slosh.shtml. Tropical Cyclone Climatology. http://www.nhc.noaa.gov/climo/. Forecast Verification. http://www.nhc.noaa.gov/verification/.

National Weather Service Directives System. http://www.nws.noaa.gov/directives/. Operations and Services. http://www.weather.gov/directives/010/010.htm. NDS 10-604 Tropical Cyclone Definitions. http://www.weather.gov/directives/sym/pd01006004curr.pdf.

National Weather Service Marine Forecasts: Coastal Warning Display Program Information. http://www.weather.gov/om/marine/cwd.htm. Rivers, Lakes, Rainfall. http://water.weather.gov/ahps/. Historical Floods in the Northeast. http://www.erh.noaa.gov/nerfc/historical/.

NOAA Central Library—Digital Documents and Maps (Weather Bureau Report of the Chief of the Weather Bureau, *Weather Bureau Topics and Personnel*, U.S. Daily Weather Maps). http://www.lib.noaa.gov/collections/imgdocmaps/index.html.

NOAA History Websites: NOAA Legacy Time Line. http://www.history.noaa.gov/legacy/time1800.html. History of the National Weather Service Forecast Office, Miami, Florida. http://www.srh.noaa.gov/mfl/?n=floridahistorypage. AOML Hurricane Research Division History. http://www.aoml.noaa.gov/hrd/hrd_sub/beginning.html.

NOAA page on Hurricane Katrina. http://www.katrina.noaa.gov/.

NOAA Photo Library—Weather Service. http://www.photolib.noaa.gov/nws/index.html.

NWS JetStream—Online School for Weather. http://www.srh.weather.gov/jetstream/. Tropical Cyclone Introduction. http://www.srh.noaa.gov/jetstream//tropics/tc.htm#origin.

S.S. Alegrete information: WreckSite—SS Alegrete (I) (+1942). http://www.wrecksite.eu/wreck.aspx?58226. Portuguese language Wikipedia—Alegrete (navio). http://pt.wikipedia.org/wiki/Alegrete_(navio).

Time.Gov Educational Resources from the National Institute of Standards and Technology. http://www.time.gov/exhibits.html. A Walk Through Time: The Evolution of Time Measurement Through the Ages. http://www.nist.gov/pml/general/time/. Daylight Saving Time—Web Exhibits. http://www.webexhibits.org/daylightsaving/.

Weather Prediction Center Tropical Cyclone Rainfall Data. http://www.wpc.ncep.noaa.gov/tropical/rain/tcrainfall.html.

INDEX